FIRE
IN
THE
EAST

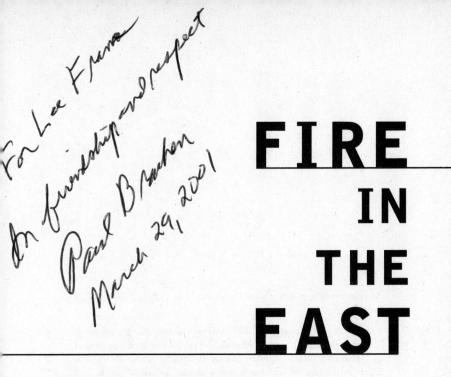

For Lee Freese
In friendship and respect

Paul Bracken
March 29, 2001

FIRE
IN
THE
EAST

THE RISE OF ASIAN

MILITARY POWER AND

THE SECOND NUCLEAR AGE

PAUL BRACKEN

 Perennial

An Imprint of HarperCollinsPublishers

First Perennial edition published 2000.

Designed by Elliott Beard

The Library of Congress has catalogued the hardcover edition as follows:

Bracken, Paul J.
    Fire in the East : the rise of Asian military power and the
    second nuclear age / Paul Bracken.—1st ed.
        p. cm.
     ISBN 0-06-019344-1
     1. Asia—Armed Forces. 2. Weapons of mass destruction.
    3. Asia—military policy. 4. World politics—1989– I. Title.
    UA830.B63   1999
    327.1'746'095—dc21   99-20103

ISBN 0-06-093155-8 (pbk.)

00 01 02 03 04 ❖/RRD 10 9 8 7 6 5 4 3 2 1

For Kathleen, James, and Meg

CONTENTS

ACKNOWLEDGMENTS

This book began in Paris in the summer of 1998. In a year marked by one crisis after another in Asia, it often seemed like the world would overtake any account of it.

Many people have come to my aid with inspiration and advice to keep my focus. I am grateful to so many for educating me in a number of ways on the issues of this book, and on much else besides. Special thanks to Jerry Adler, Patti Benner Antsen, Seth Carus, Bob Cassidy, Lewis Dunn, Tom Evans, Mickey Goodman, Robby Harris, Charles Herzfeld, Tom Hirschfeld, Ben Huberman, Paul Kennedy, the late Paul Kreisberg, Andy Marshall, Bill Overholt, Chae-ha Pak, John Petrie, David Rich, Mike Salomone, Michael Sherman, and Martin Shubik. I would also like to thank my students at Yale, from whom I've learned more than they will know.

Special thanks to Kim Witherspoon for her constant encouragement and support. At HarperCollins David Hirshey, Cass Canfield, Jr., Jay Papasan, Tim Duggan, and Peter Shea have been extraordinarily encouraging and helpful. Finally, my biggest thanks goes to Nan, who put up with this project no matter what the demands were.

INTRODUCTION

For two hundred years the world has been shaped by the fact of Western military dominance. Gunboats as agents of national power have been supplanted by warplanes, and they in turn by missiles and satellites and computers, but until very recently all were a monopoly of Europeans and North Americans. Now that monopoly is coming to an end. Missiles carrying atomic and biological warheads will, within the decade, be within the reach of as many as twelve Asian nations, from Israel to North Korea. The world the West has known—constructed, in part, for its own convenience—will change enormously. Whether the change will also be catastrophic is the subject of this book.

The West's military dominance has held for so long that it is taken for granted, the unnoticed backdrop to international affairs. No one even thinks about a world without it. But a world of new military powers is appearing right before our eyes. Iran fires long-range missiles and is very likely to test a nuclear weapon in the next few years. North Korea, already believed to have a few atomic bombs, shoots a missile over Japanese cities with a range that could strike U.S. territory. India and Pakistan are building serious nuclear forces—not experimental devices for status, but weapons akin to what the United States and the Soviet Union had in the 1950s. Washington and its allies used every tool of diplomacy to stop their atomic tests in 1998. Their answer, essentially, was that countries with atom bombs don't have to listen to anyone else.

Asia's new military might was already a major factor in international politics. After confrontations with China over human rights and trade in the early 1990s, the United States reversed course following Chinese missile tests near Taiwan in 1995 and 1996. Washington showered attention on Beijing, giving its leaders a twenty-one-gun salute and overlooking continued violations in human rights and the sale of missile parts to Iran and Pakistan. What happened to make the United States change its mind? Officials explained that engagement with China was better than containment. This is true. But the reason for the diplomatic shift was military power. The widespread sense that China will be a major military power makes an arrangement with it an American necessity. The United States was not as concerned about "engagement" with China in 1965, when its armed forces consisted of 5 million peasants with rifles and two atomic bombs.

India drew this same conclusion. It was the major reason behind its nuclear breakout in 1998. India saw the Chinese-American partnership working against it. Delhi correctly foresaw that the Western economic sanctions imposed following the tests wouldn't last long because, after all, it was better to "engage" India than to "contain" it. Iran is drawing this lesson as well, building up long-range missiles and atomic bombs while making diplomatic overtures to the West. Will Washington elect to "contain" Tehran—or to "engage" it?

Atomic bombs on missiles get the West's attention. If the country is big and powerful, it earns Western engagement. First come the international Ping-Pong games or wrestling matches, building goodwill through sports. Next, foreign investment is encouraged.

Technology is released for sale, and before long supercomputers—which can be used to design nuclear warheads—are on their way.

But criticizing Western policy isn't the point of this book. Engagement may be the only sensible diplomatic solution for dealing with countries with nuclear weapons and missiles. Criticizing that policy is like spitting into the wind. The problem is that the United States isn't thinking about what it will be like to live in a world where five to ten Asian countries are nuclear powers, with missiles that can hit distant targets. Instead, it's focused on keeping this from happening, despite evidence that the policies that have worked for the last twenty-five years are rapidly losing effectiveness. Proliferation of modern weaponry is driven not by anything that happens in Washington, but by the national strategies set in Beijing, Delhi, and Tehran. The spread of these technologies will continue, and as it does so, it is likely to become self-reinforcing. As China, India, and Iran get these weapons, North Korea, Pakistan, Iraq, and Syria have a much greater incentive to follow. Israel will be watching these developments closely. All these countries will have a much greater ability to ignore any embargo of Western technology or arms control roadblocks meant to slow them down. This is the new political reality the West will face in the next century, and major challenges to U.S. national security and international order will result from it.

First, the West's self-conception as the architect and maintainer of international security will change as Western military influence shrinks in Asia. The United States will be a lot more cautious in dealing with countries that can strike back, either at its bases and outposts or directly at the American homeland. Since the War of 1812, only one country in modern history has ever been able to mount a convincing threat to the territory of the United States—the Soviet Union. Now there will be many.

For U.S. allies in Asia, the dangers are much greater. For Israel and the Gulf states, they involve survival. Historically, Israel confronted low-tech opponents who couldn't reach it except by overland attacks with big armies. Its air force and army could defend against those. In the Persian Gulf, small American contingents could deter attack. Now the balance of force has been turned upside down. Iran, Syria, and Iraq (if it ever escapes United Nations sanctions and scrutiny) will be able to strike with missiles carrying atomic or chemical bombs. Israel will have to restructure its armed forces entirely, placing far more reliance on nuclear weapons and other deterrents and less on tanks and airplanes.

Any new security order in Asia will face some fundamental dilemmas. The countries that control most of the world's oil supplies are weak and unlikely to be able to defend themselves against stronger neighbors. New oil fields in the trans-Caspian region or in Russian Siberia could offset dependence on Persian Gulf oil, but they are even harder to reach from the United States than the Gulf. Japan's dependence on the United States for its defense has been a settled question for fifty years. Now the underlying assumptions behind it are changed, and the question will be raised again. The states of Southeast Asia will have to find a place in this new environment as well.

For the first time in history, Asian states can attack one another's homelands. The barriers of distance and harsh terrain that have separated enemies for centuries and worked to make regional politics relatively stable are disappearing as new arms extend military reach far beyond that of slow-moving peasant armies. Beyond all the uncertain predictions of what China may or may not want in the future, the debate over whether Iran is really becoming more moderate, the explosive potential of the Kashmir dispute, and all the dozens of Middle East peace proposals and conferences, one overarching change has gone unnoticed. Since the 1940s, when the West directly controlled the Asian security system, Asia has been as stable as it has been because the primitive character of its armed forces made them blunt political instruments outside of their own territory. Now they can get at one another. More than anything else, weapons of mass destruction use intimidation and threat for their effect. As in the first nuclear age, brandishing them for political uses is their most potent effect. It is this political purpose that is so troubling. Whether Asia, and the world, can contain the international dynamics unleashed by weapons of mass destruction will be the other great challenge of the twenty-first century.

THE POST–VASCO DA GAMA ERA

There are many different ways of seeing the world. After the collapse of the Soviet Union, the world was said to have entered a post–cold war era. While this label did not convey much information, it did declare that the most important historical event of our era was, in fact, the cold war. This notion is supported by only the most parochial reading of history. The cold war shaped Europe much more than it did Asia.

In the most basic terms, the cold war originated with a Russian tank threat to Western Europe. The whole confrontation started when Stalin kept his army forward in Europe in 1945, within striking distance of France and West Germany. Over time the superpowers expanded the terms of this somewhat confined showdown to the far corners of the globe. But thus transplanted, the cold war was transformed into a very different kind of struggle. The most important problem for Asia after World War II was ejecting colonial powers and establishing stable nation-states. Later its challenges were industrialization and economic growth. To speak of the cold war in Asia is to focus not on what was most important but on the West's egotistical assumptions about the importance of a struggle it tried to impose on the world. The United States learned this the hard way in Vietnam, when its cold war strategy for Europe—containment—was blindly transplanted to the very different conditions of Asia.

Of course, China and India played roles in the cold war, but not the roles that the United States—or for that matter, the Soviet Union—had chosen for them. Both exploited the cold war as a cover for more important matters, as a way to extract payments alternately from one superpower and then the other, as many other Asian countries did. Their struggle was never to expand or resist communism. Rather, it was to build nation-states with enough stability to industrialize and modernize. In so doing, China and India unleashed powerful forces that are transforming a continent over four times the size of Europe.

For Europe and the United States, though, the end of the cold war was said to have brought history back to year zero. The image of the Berlin Wall coming down in 1989, and the subsequent economic boom, created a triumphal aura about the West's innate superiority, bordering on the delusional. One popular television advertisement meant to assure investors during the Asian economic crisis even announced that investors should relax and stay in the stock market for the long haul because it was only "year nine" after the crumbling of the Berlin Wall. The world was only nine years old. The past simply disappeared.

The post–cold war era never happened. It was a Western conceit, imposing its view on the world and artificially linking the end of the Soviet Union to the economic boom that had been under way for decades. The industrialization and economic growth in Asia in the 1990s started in the 1960s, when Japan proved that it could be done, that industrialization was not a Western monopoly. This had nothing to do with the start or end of the cold war. Rather, it proved that what Japan could do, so could China and India. If it could be done in industry, it could also be done in the military. The world now is going through this more basic transition, full bore into what is not a post–cold war era but a post–Vasco da Gama one. It is a world where Asian economies and Asian military power are much more important factors in world politics, and where the automatic presumption of Western control over each no longer holds. Seen this way, the Asian financial crisis of the late 1990s and the spread of new weapons are connected. Both stem from the industrial transformation of Asia that has opened up new technological and economic futures, positive and negative.

The year 1998 was significant in this regard for something that didn't happen, a non-event that shows how differently the West and the East measure history. May 1998, the month of the Indian atomic explosions, marked the five-hundredth anniversary of the first landing of the West by sea in Asia. Vasco da Gama's voyage from Portugal around the Cape of Africa to India in May 1498 marked the first encroachment by the West into the East using the high technology of the day, the ocean sailing ship.

In its human and technical accomplishment, Vasco da Gama's voyage ranked with the landing of Neil Armstrong on the moon. But 1998 saw no gala celebrations of da Gama and his feat. It was an anniversary that the West conveniently forgot. Clumsy efforts by Lisbon to commemorate the event animated more antipathy than empathy in Europe and were seen by India and other Asian countries as an affront. At the Sorbonne a small exhibit displayed some treasure boxes and cannons typical of the voyage. Unintentionally, this exhibit captured the reason there was no greater acclaim for the expedition: the items on display symbolized how its larger scope is seen in Asia. The forcible opening of Asia by the West ushered in five centuries of arrogant colonial rule, piracy, slave trading, and plunder on a scale previously unknown in history.

The Columbian five-hundredth anniversary observance in 1992

produced a similar discomfort in the United States. In great American universities, the Columbian expedition could not even be referred to as the discovery of America. Rather, it was the encounter between the Old World, Europe, and the New, one with devastating consequences for the native culture. At the end of the twentieth century, the da Gama and Columbian voyages inspired either indifference or embarrassment in the West.

But there was a crucial difference between the two encounters. European civilization wiped out what it found in North America. Native American civilization was totally destroyed and replaced by that of the newcomers. In Asia this didn't happen. Western influence, from the Near East to the Far East, was never more than a thin overlay on what had always been there. Indeed, in Asia it couldn't happen. Civilization in Asia was, and is, so rich and so developed that the West could never replace it, despite centuries of economic and military dominance.

China, India, and Persia were once the great civilizations of the world. Their positions were lost when the West, with its superior commercial and technical skills, took colonial mastery in Asia. In the twentieth century, using conventional measures of development like steel production, telephones per thousand population, and number of automobiles, these civilizations fell even further behind.

What is now taking place is a reassertion of these societies, with a capitalist economy and the modern technology of atomic bombs and missiles. Seen this way, the end of the Vasco da Gama era has the most profound implications for the world, far more than the end of a Communist state established in 1917.

Western colonial rule in Asia formally ended after World War II. India and Pakistan gained independence in 1947, and the British departed from South Asia. The British left China in 1949, fleeing Nanking one step ahead of Mao's army. By the 1960s Britain had pulled back from virtually all of its earlier outposts, from Malaya to

Suez. The lowering of the Union Jack over Hong Kong in 1997 marked the end of British dominion in Asia.

The French pulled out of Vietnam after their defeat at Dienbienphu in 1954, handing the problem over to the United States. The strategy at the time was for the United States to replace the European powers and halt the spread of communism, not as a colonial power but by treaties, using military and economic aid to make the region stable and prosperous.

Communism extended the Western stay in Asia by forcing the United States to become the new guardian of international order there. Out of the postcolonial wars that convulsed Asia from the 1940s to the 1960s emerged a number of extraordinarily successful and stable nation-states. With the help of the United States, most of them have weathered domestic turmoil and undermining from Moscow and Beijing.

But the departure of the West from Asia continued throughout the period. It is important to understand this. Asian societies tend to look at political events over a longer time frame than the West. Consider two events that, from the American perspective, seem completely unrelated: the withdrawal from Vietnam after the Communist victory in 1975, and the closing of the big U.S. bases at Subic Bay and Clark Airfield in the Philippines in 1992. The evacuation from Saigon was the outcome of a failed military effort to beat back a Communist insurgency. Washington views the Vietnam War in the solipsistic context of American anticommunism, the domestic upheavals of the 1960s, and the careers of various American politicians, but not as part of a larger historical or geopolitical trend. The withdrawal from the Philippines, by contrast, was peaceful and, as viewed from Washington, the result of a regrettable national disaster and an understandable dispute over lease payments.

But seen from Asia, the two events are part of the same ongoing phenomenon. Nanking, Dienbienphu, Saigon, Suez, the Philippines, and Hong Kong were all Western outposts and centers of military presence. Now they are gone for the West. The trend begun in 1947, of Asia becoming Asian, continues; it goes on through Vietnam Wars, Philippine base closures, the economic boom of the 1980s and 1990s, and the financial crisis of the late 1990s. Viewed this way, the United States is following exactly the path of the European colonial powers, vowing to remain forever but each year pulling back a bit more.

The United States still holds bases in the Middle East, South Korea, and Japan, but for how long, at what cost, and under what conditions are perennial questions. The United States may remain in Asia. Countries in the region in many ways need American alliances now more than ever for their own security. But the broader conditions of staying are changing fast. In particular, the era when the greatest military power in Asia was not an Asian country at all is coming to a rapid end. Staying will mean something very different than it did in the past.

THE NEEDHAM PARADOX

If viewing the cold war as the fulcrum of history is a Western conceit, then so is imagining that the West's current lead in technology and innovation constitutes a permanent natural advantage. Even leaving Japan aside, it is now well known that the Chinese can make computers and that Indians can write programs for them. But industrialization in Asia has implications that go far beyond these countries' balance of trade. These industries are signs that a knowledge and innovation infrastructure is flourishing in Asia. Industrial capitalism makes societies more dynamic than agricultural ones. Changing social roles, new patterns of communication, and a new openness to innovation represent a societal transformation of kind as well as degree.

The work of the great China scholar Joseph Needham makes a very telling point about the exploitation of innovation. Needham wrote a comprehensive history of Chinese science and technology. He discovered that most of the inventions that transformed the West were originally developed in China. The printing press, gunpowder, and the magnetic compass all appeared in China before Europeans had them. Between 1000 and 1500 A.D., China was the most advanced society in the world. It was also the greatest military power. One hundred years before the Europeans landed in India, China had built the most advanced navy, one far larger than any such fleet in Europe at the time. In 1281 China tried to invade Japan with a fleet of 4,400 ships. Its navy sailed to the east coast of Africa and conducted vigorous trade over the Indian Ocean. In the early 1400s the largest Chinese ships displaced some 1,500 tons, compared to Vasco da Gama's flagship of 300 tons. By any measure—ship size, numbers, navigation—the Chinese fleets were better than anything in Europe.

Yet in a few years, beginning in 1428, shortly before Europe started on its path as a world power, all of this tremendous military capability went into steep decline. Chinese ships were brought home. Shipbuilding was outlawed. China became inward-looking.

This development led to what is known as the Needham Paradox. Why did China not remain a great power, given its huge lead in technology over the West? The answer Needham offered is that, far more important than anything the West did, the Chinese bureaucratic and feudal system worked to suppress innovation and preserve a system that had little interest in the outside world. In particular, China suffered from the absence of a powerful business class to offset the stifling power of the central government. Narrow bureaucratic interests stopped the spread of China's naval power. A decline in maritime-related technologies soon followed.

The Middle East also once had a clear advantage over the West. As late as the sixteenth century the Ottoman Empire was militarily far stronger than any power in Europe. Had the empire won certain key battles in the seventeenth century, Europe might never have developed to become the center of world power that it did. France would have been a mere province of the Ottoman Empire.

What Asia's industrialization signifies is the end of the conditions of the Needham Paradox. Once, government smothered innovation. Now, almost everywhere in Asia, governments are encouraging it. One Asian country after another is following the path blazed by Japan, the first Asian country to industrialize. It is true that this process is not as developed as in the United States, and that it has tremendous inefficiencies. Crony capitalism is alive and well in Asia. But the key word here is *capitalism*. Crony capitalism is much more dynamic than agricultural communism. As Asia moves into market economies, its potential for innovation sharply increases.

And even though modern capitalism was a Western invention, there's no reason to think Westerners are the only ones who can master it. To the contrary, Herman Kahn has argued that for cultural reasons capitalism is more natural to Asia than it ever was to Europe. The spirit of devotion to one's task that is the hallmark of a productive working class had to be painstakingly inculcated into a mostly balky peasantry in Europe, even while it flourished under Confucianism. Even today the work ethic in China puts Germany to shame.

The energies that capitalism awakens can be channeled in many different directions—from mass consumption to technology to military power. The West's expectation is that all this dynamism will go into peaceful commercial pursuits, but it is almost certainly mistaken. Saddam Hussein showed this with his mini–Manhattan Project, a sprawling nuclear weapons complex linking multinational suppliers, front companies, research laboratories, universities, and imported technical talent. The West was greatly surprised at the scale of Iraq's arms programs not because of inadequate intelligence gathering but because of its mindset: a project of this scope and complexity was supposed to be beyond the ability of a Middle Eastern nation whose entire gross national product was one-sixth of General Electric's annual revenues. If it continues thinking this way, the West will be hit with one surprise after another until it understands that the fundamental conditions of Asian backwardness have changed, and that the dynamism of Asian societies is a new resource that is not yet picked up in GNP statistics or other financial measures.

TROUBLE ON THE ROAD TO McWORLD

The future of Asia is often described as a battle between globalization and backward nationalism. *Globalization*, the international linkage of economies and cultures, is supposed to create a shared destiny for the planet that will be a major force for peace and stability. Multinational corporations, worldwide television networks, booming stock markets, and an equally vibrant and expanding Internet are all manifestations of globalization. By this theory, the sale of technology across national borders is a potent force for peace. The more these things spread, the safer the world will be. Each new satellite dish and McDonald's helps build a new middle class that more tightly links the world together.

Nationalism is the straw man that globalization is intended to overcome. It connotes religious extremism and fundamentalism, ethnic chauvinism, hostility to change, and opposition to all the democratic values so cherished by the West. Where globalization brings people and economic interests together, nationalism divides them. Its turf is the rural village where elders are threatened by young people listening to Michael Jackson; its struggles are tedious and pointless military confrontations in the Middle East and Kashmir; its symbol, increasingly, is the Islamic religion itself.

The unquestioned assumption in the West is that globalization is the force of the future, while nationalism stands for the world of the past. Seen this way, the long-term outlook is comforting, since the future by definition always arrives eventually; the only question is whether in the next few decades Western, democratic values will spread fast enough to persuade Asian countries to give up missiles and atom bombs for video games and hamburgers. This is a very pleasant view of the world, defined in clear and simple terms.

Globalization does exist, and it is a very powerful force. Whole new interest groups have appeared in the world: Chinese entrepreneurs, Indian software programmers, Thai stockbrokers. Globalization surely describes the social and economic trends of the last few years, with the world economic boom, the opening of international markets, and the decline of tyrannies, from Berlin to Moscow.

Or does it? Western analysts love simple descriptions. Even the most complex problem is mapped onto a binary choice. China is either a friend or a foe. India can choose guns or butter. Globalization will either triumph or succumb to backward nationalism. But these simple binary choices fail to come to grips with realities that are slightly more complex.

The globalization debate is not even new. In the nineteenth and early twentieth centuries the greatest social thinkers the Western world has ever produced had a very similar prognosis for their world. Karl Marx and Max Weber both believed that the most important force of their day was industrialization. Marx and Weber made a strong case, correctly as it turned out, that the future would be characterized by the internationalization of industry. Marx argued that it would create a consciousness among newly empowered groups—the organized proletariat—that would press their interests without regard to national borders and smash the old empires of Europe.

Weber argued that it would lead to the standardization of social and working life to create a common culture around the globe. For him, the large bureaucracy was so efficient compared to the small craftsman that it would inevitably spread across the globe. Weber recognized the power of good organization. The worldwide reach of McDonald's, the Internet, and stock markets would not have surprised him in the least.

What would have given both Marx and Weber a great shock, however, was the extraordinary violence of the first half of the twentieth century. Both believed industrialization would overcome national boundaries to become a new pacific force in the world, a force that would bring countries together rather than rip them apart. In some ways they were right about this, too. Factories, markets, and even the communications systems of the day did exert a strong unifying and standardizing force. Industry and society in different countries took on similar characteristics. The factory in England looked like the one in Germany. The organization of work, housing, and education created similar problems in all the different industrializing countries.

But Marx and Weber greatly underestimated the strong appeal of nationalism, as well as the power of nation-states to manipulate the globalizing tendencies for their own purposes. Nationalism was a response to change, but it harnessed the very energies that change had unleashed. As the pull of globalization heightened the necessity of internal political cohesion, it paved the way for nationalism. Marx and Weber also overlooked one implication of their argument that should have been obvious. Industrialization made war a far more deadly affair than before, because the same technologies that gave rise to the sewing machine could be applied, with sufficient ingenuity, to making a machine gun.

Not much is being said today about the place of nationalism and military power in the organization of Asia; it is much more popular and agreeable to focus on the stabilizing and pacifying effects of globalization. It is strange, however, that the role of nationalism and military power is being overlooked, because in the Western experience the arguments have all been made before, by Marx and Weber, and have been made much better than they are now.

The recent financial crisis in Asia provides a fascinating illustration. The dangerous scenario evoked has to do with unleashing the forces of nationalism, which in turn could promote protectionism, and somehow this could lead to military aggression, although the connection to warfare is never specified. The dark forces of nationalism, protectionism, and militarism are grouped together. The forces of enlightenment—globalization, the Internet, and free trade—are lined up on the other side. It is a Manichean battle for the future of the world, at least as it is constructed in the minds of Western intellectuals.

But the disturbing military events of recent years, the missile and bomb tests, the biological warfare programs, and the chemical weapons were the products of a prosperous, liberalizing Asia. None of these programs was a sudden negative response to financial crisis. It takes many years to build these weapons. Research must be carried out, experts hired, and weapons fabricated. Only after all of this is done can they be tested. The Indian atomic bomb decision was probably made sometime in 1987. The Chinese missile program has been under way for many years. Even Iran's new missile capabilities were developed after its war with Iraq, when it became rich enough to pay for them.

These weapons programs had been under way not only for many years but at a time of widespread prosperity and (at least in China and India) market liberalization and growing acceptance of many aspects of Western culture. For some reason it always comes as a shock to Americans that people who eat Kentucky Fried Chicken and watch MTV may not share American values on a deeper level.

Moreover, in every case, from Israel to North Korea and countries in between, the forces of globalization have contributed to weapons programs. Critical parts were acquired without much difficulty, as Iraq and Pakistan could simply turn from the unwilling United States to much more cooperative France or China. Saddam Hussein's electronic warfare and intelligence systems were bought on the commercial market in Japan and France. Iraq's bomb parts were smuggled through front companies operating in Switzerland. And its biological warfare starter kits were bought in the United States, from a federal government bureaucracy interested in the global spread of scientific knowledge.

Again, Max Weber was right. The Iraqi bomb program looks a lot like the Indian, Pakistani, Israeli, and Chinese programs. How could it not? As he foresaw, the modern bureaucratic way of putting things together is far superior to old-fashioned craftsmanship. This is as true for atomic bombs as for anything else.

The Asian financial crisis was disturbing because it was seen as a raging fire that could jeopardize the central achievements of American foreign and economic policy in the 1990s: the stabilization and prosperity of East Asia and the movement toward an open world economy. But with its focus on economics, the West has failed to recognize that technologies of wealth and war have always been closely connected. Without understanding this basic relationship, it has been too easy to see militarization in Asia as the product of a few rogue states that defy the forces of globalization—and whose people therefore pay the price, in poverty and political oppression. All of these observations are sometimes true, and no one would defend the North Korean way of life or the lunacy of its leadership.

But left out of this analysis is the simple point that the military leap forward to weapons of mass destruction and missiles is not confined to the rogue states. In Asia it includes a much broader set of countries—Israel, Syria, Iran, India, Pakistan, and China. Judging by the standard of expanded weapons of mass destruction, there are now eight "rogue" states, and this does not even count Saudi Arabia and Vietnam, both of which have long-range missiles.

Industrialization and globalization increase military potential. That is the record of the 1990s. To take two wildly opposite cases, both Israel and North Korea, one a democracy and one a totalitarian dictatorship, have exploited their opportunities to build weapons of mass destruction. The difference between today and the year 1900 is not that the world is more interdependent than it was then. By many measures—foreign direct investment, currency convertibility—that world was more interlinked than it is today. The difference between the two eras is much more prosaic. Then a country had to build its own military technology. Germany couldn't buy its battleships from England. It had to rely on its own industrial power. Now countries can buy almost whatever they want from others, using international markets greatly abetted by the forces of globalization. Need zirconium fuel rods for a uranium enrichment plant to make a nuclear bomb's exploding core? No problem. At least a dozen countries offer these, no questions asked. Only through globalization could an Iraq or a North Korea, two backward, impoverished countries that started twenty years ago with extraordinarily primitive technical establishments, develop their weapons of mass destruction as far and as quickly as they have.

Globalization does not make the world more dangerous per se. It does not cause war any more than industrialization caused war. Rather, globalization, like industrialization, is not a political force at all. It is an economic and social one.

Many of Asia's atomic bombs were built over the objection of key groups. In Israel most scientists opposed the bomb from its beginnings. In North Korea the average citizen was likely to be far more interested in getting something to eat than in developing a national bomb. But this didn't matter. What mattered was that governments used their power to go around these and other obstacles. The same government power used to industrialize was also used to militarize. That was the problem in 1900, and it is the problem in the year 2000.

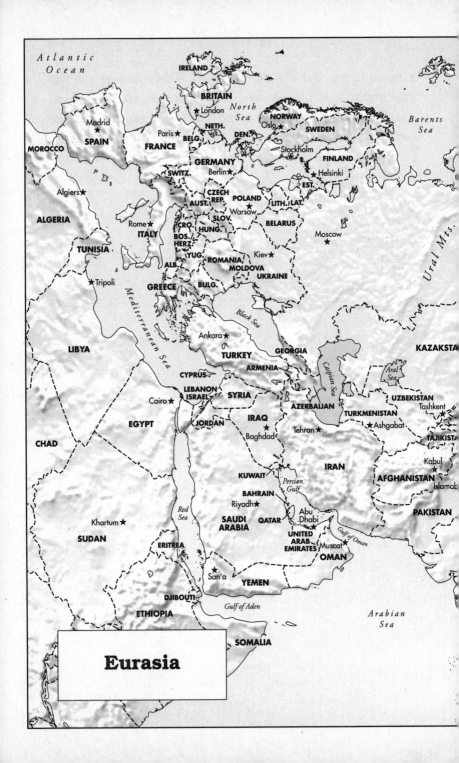

Eurasia

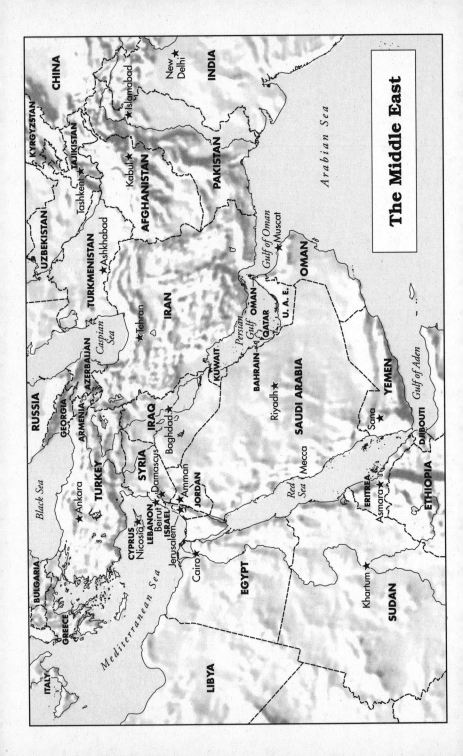

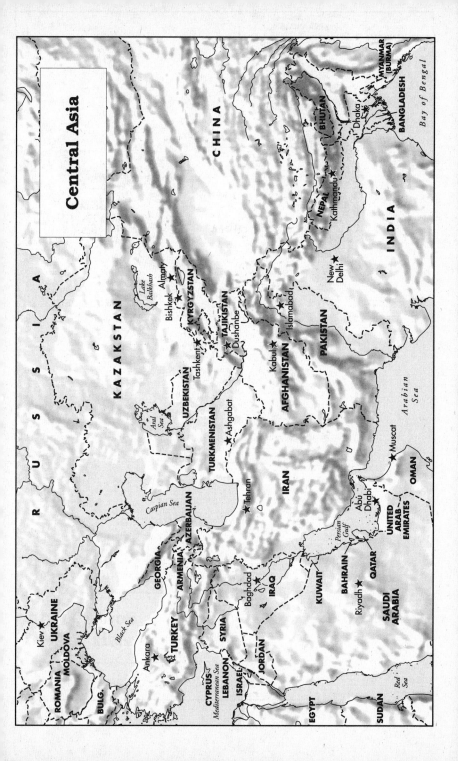

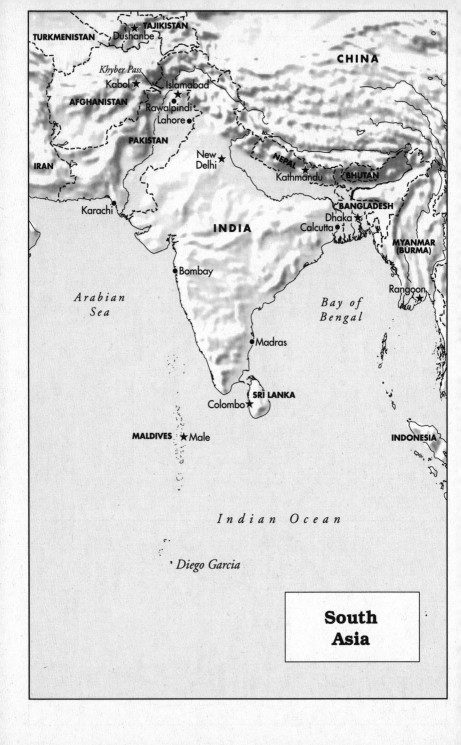

South
Asia

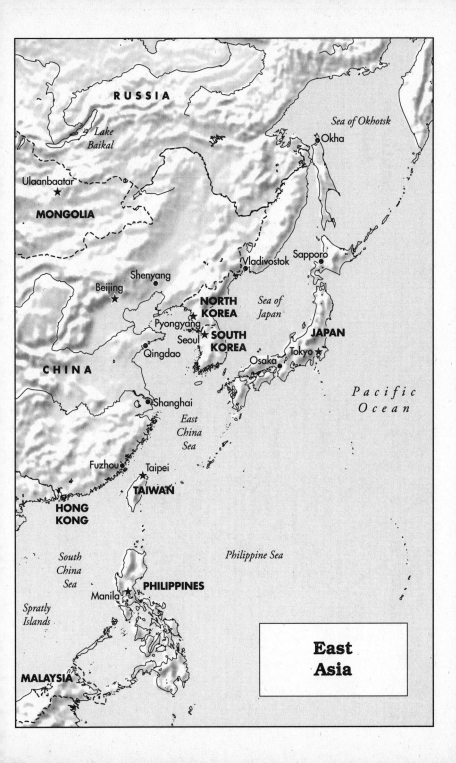

RUSSIA

Sea of Okhotsk

Lake
Baikal

Okha

Ulaanbaatar ★

MONGOLIA

Vladivostok

Sapporo

Shenyang

Beijing ●

NORTH
KOREA

Sea of
Japan

Pyongyang ★

SOUTH
KOREA

JAPAN

Seoul ★

Qingdao ●

Osaka

Tokyo ★

CHINA

Pacific
Ocean

Shanghai ●

East
China
Sea

Fuzhou ● Taipei ●

TAIWAN

HONG
KONG

Philippine Sea

South
China
Sea

PHILIPPINES

Manila ★

Spratly
Islands

MALAYSIA

East
Asia

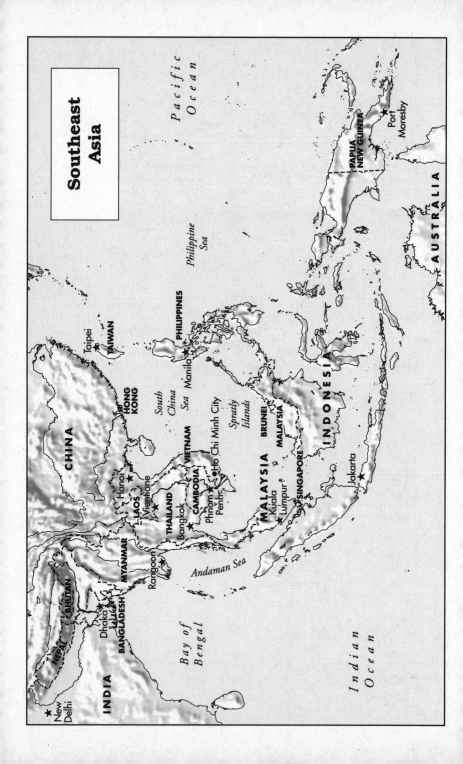

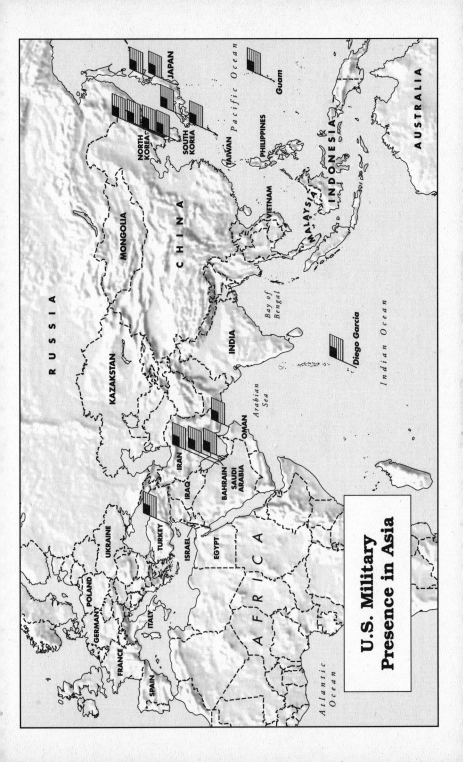

U.S. Military
Presence in Asia

1   NO ROOM ON THE CHESSBOARD

The world is moving at warp speed. A button pushed at a trading desk in New York affects prices around the world in seconds and ripples through the world's economy in a matter of days or weeks. Transfixed by twenty-four-hour news broadcasts and by real-time financial data around the clock and the mountains of information flickering continuously across the Internet, Western leaders in the 1990s have devoted themselves to detecting and responding to short-term phenomena.

But the world is also moving at slow speed. Slow-motion change is barely perceived. When India and Pakistan tested their atomic bombs in 1998, Western leaders were transfixed by a stock market collapse in Indonesia. The twin bomb programs had been under way for fifteen years, but Western leaders, and certainly the media, were absorbed in the breaking story—a financial panic! hurried conferences of central bankers aimed at restoring confidence! statements! leaks! denials!—right up until the video of the blasts showed up on CNN. Only then did the nuclear arming of South Asia, overlooked for years, commandeer the world's attention.

In *Slowness*, the novelist Milan Kundera draws a connection between change and forgetfulness. We are caught up in the spiral of events, lost in its energy, blind to the accumulation of slow changes remaking our world. Without our noticing, the political and military map of Asia—one-third of the earth's landmass, with almost two-thirds of the world's population—is being redrawn. The Asia of the cold war, a disjointed collection of subregions and military theaters, no longer exists, not even notionally. Instead, the West must adopt a new paradigm, a geography of strategic interactions, in which the old barriers of distance and terrain have lost their meaning. These are some of the factors shaping it:

- Europe, called the cockpit of the world because it has been the locus of so many major wars, is now more secure than it has been in ages. As a result, European armed forces have been cut back to the point where Europe is no longer a serious military power. The British navy takes to the seas with centuries of proud tradition behind it, but with fewer submarines than India has. The French armed forces are so technically backward that they are virtually irrelevant except for low-intensity peace-keeping missions. European armed forces are hopelessly unprepared when it comes to the kind of modern fighting the United States engaged in during the Gulf War. Between nuclear retaliation and peacekeeping, they have few capacities.

- An unbroken belt of countries from Israel to North Korea (including Syria, Iraq, Iran, Pakistan, India, and China) has assembled either nuclear or chemical arsenals and is developing ballistic missiles. A multipolar balance of terror stretches over a 6,000-mile arc, comprising some of the most unstable countries on earth, with no Western allies in the sense in which the term is used within the Atlantic alliance.

- This arc of terror cuts across the military and political theaters into which the West conveniently divided Asia, essentially for the purposes of fighting the cold war: the Middle East, South Asia, Southeast Asia, and Northeast Asia. The ballistic missile, once launched, does not turn back at the line that separates the territory of one State Department desk from another. Thus, the Gulf War brought the troubles of the Persian Gulf to Israel, linking theaters that had once been considered separate. Israel, for its part, sends up satellites to spy on Pakistan, 2,000 miles away, spooking Islamabad into seeing an Indian-Israel squeeze play against it. Chinese and Indian military establishments plot against each other, making East and South Asia one military space.

- The great interior of Asia is in play again. What used to be called "inner Asia" was stable for most of the century in the iron grip of two Communist giants. But in the face of Russia's decline, China, Iran, Pakistan, and Turkey are all vying for influence among the Central Asian countries that emerged from the Soviet breakup. These are countries that even most educated and well-traveled Americans know virtually nothing about: Kyrgystan, Tajikistan, Uzbekistan, Turkmenistan, Kazakhstan.

- The world energy map is being redrawn as China, India, South Korea, and Southeast Asia industrialize. New oil and gas fields in the trans-Caspian, in Central Asia, and beneath the Asian continental shelf will radically change the direction of flows of oil and gas around the globe.

Despite these profound changes, the United States continues to see Eurasia as a chessboard, where the object of the game is to prevent the rise of any country that could challenge Western military superiority. The West's dual strategy is to pursue its own technological superiority at all costs while trying to keep any other player from amassing advanced armaments. The United States and the Soviet Union, even as they competed in the cold war, essentially ran things so that no third country could upset the Asian balance. That was easy when Asian military capability was limited. But year by year the playing surface is shrinking and the game is changing as the pieces on the board become more powerful. The ballistic missile has empowered pawns to check the dominant powers; countries that were once pawns now have the reach of knights and bishops.

In the early 1990s the United States pretty much ran things by itself. But in the face of increasing Asian missile power and weapons of mass destruction, it has tried to include China as a possible partner in imposing political and military order on the continent. The Indians, not caring for this arrangement, made their views known with five nuclear shots. Much more than a military test, the 1998 Indian exercise was a signal from one of the pawns about how the game was going to be played from now on. Players on the Eurasian chessboard are running out of room. The maneuver space is becoming more tightly coupled. When a move is made in one area—U.S. partnership with China—it reverberates almost at once in another—India testing nuclear bombs to signal disapproval. New rules will have to be written, and new strategies developed, to replace the "arrangement politics" that prevailed when one or two major pieces dominated a board filled with pawns.

ASIA'S CHANGING GEOGRAPHY

The chessboard is also a map, and looking at it this way is useful: it forces us to think through in concrete terms how physical units—armies, ships, satellites, missiles—can be deployed in space and time, and how political power grows from them.

It was European sea power that defined Asia as a political and economic entity in the first place. Before the European maritime age of the sixteenth century, *Asia* was not a unitary concept. For the West and its map makers, it was a fabulous region of exotic spices, fabrics, customs, and mythic monsters, but in one sense they understood the continent better than their successors did: just as we do today, they grasped the immense diversity of the place, the incongruity of grouping together lands as disparate as the Arabian peninsula and the Korean.

But in the sixteenth century European nations developed the technology of ocean navigation, using heavy ships guided by the compass, and one after another they were able to build empires far from home. Asia thus took on a politico-economic unity in the Western mind: it was a place for colonies and for trade. This great paradigm shift in turn made new demands on European sea power. Ships had to be built to withstand the rigors of the long voyage around the African Cape. Fleets were sized according to the colonial ambitions of the throne. A new kind of political structure evolved, the overseas maritime empire; lasting until the middle of the twentieth century, it created a pattern of international politics that endures down to the present time. For Asia that pattern has enabled outside Western powers to draw the maps and write the rules of the game—in setting economic policy, setting the rules for joining the Western club, and in trying to halt the spread of nuclear weapons.

Western powers backed their decisions with military authority. In the colonial era Europe and Russia were the dominant powers; after World War II it was the United States and the Soviet Union. In all cases the dominant power was a Western country, extending its power across vast distances of sea and land. The Europeans used galleons and battleships; Russia employed railroads. The United States has deployed aircraft carriers, land-based bombers, satellites, and cruise missiles. But the spread of missiles and weapons of mass destruction to Asian nations signals the end of even American dominance.

There is no accepted term or field of study to describe how changes in military technology lead to such changes in geographic influence, but perhaps *military geography* will do. The term is not restricted to the way terrain affects army movements. As used here, military geography also refers to the way developments in weapons, communications, and transportation affect larger political arrangements—like empires, the cold war, the "Pacific Basin," and a U.S.-Chinese understanding about the future of Asia. All of these are political constructs that have changed slowly, until the technological conditions that gave rise to them changed, to the advantage or disadvantage of different actors. Military geography is not the same thing as geopolitics—the strategic management of geography for political purposes.

When the United States guarantees the security of South Korea and Japan, it is using geography for political purposes. The U.S. goal is to prevent another power from taking over these allies, either by direct occupation or by intimidation. A country that succeeded would control such a large population, GNP, and technology base that it could endanger the security of the United States.

Looked at geopolitically from the U.S. perspective, Asia is a vast continent full of strategic management challenges. In Southwest Asia, the world's largest supply of petroleum sits near countries that would like to control more of it—thus Iraq's seizure of Kuwait—or that would use it to finance their war against Israel. In the Far East, two U.S. allies, Japan and South Korea, face a threat from an erratic state, North Korea, and the rising influence of China. These two fronts, in Southwest and Northeast Asia, have consumed most of the military and diplomatic attention paid to the continent by the United States since the end of the cold war.

The shape and terrain of the Asian chessboard constrains what the United States can do to shape events there. Eurasia—the combination of Europe and Asia—is a vast crescent, bounded on the north by the Arctic Ocean and by Africa, the Indian Ocean, and the Pacific on the south. To the west lies Europe, and to the east, Japan. The heartland of this immense continent—a rugged topography of mountains, deserts, and frigid wastelands—confines outside actors to the crescent's rim. No Western power has ever succeeded in operating anywhere but in these coastal states. Today the United States relies on Japan, South Korea, and Saudi Arabia as the base of its military power in Asia. Without them, the distances to project power in Asia are too great, the climate and terrain too extreme. What has been happening is that many of these coastal states—Iraq, Iran, Pakistan, India—are no longer open as bases for Western interests, as they were in earlier times. As the United States is pushed to a much thinner rimland from which to operate, we see the West being forced out of Asia.

Thus, U.S. geopolitical strategy in Asia today stands in a long tradition of Westerners using the rim of the continent to block or moderate the influence of larger continental powers. This was the case in the cold war, when South Korea and Vietnam were the venues to block Chinese influence. The weak desert kingdoms of Saudi Arabia, Kuwait, and the Gulf states—also on the periphery of the continent—were the agents for checking Iraq and Iran. And whenever possible, the United States tries to exploit divisions between the Asian powers—notably, by supporting Iraq in its war with Iran.

This is a centuries-old game, one at which Britain excelled for much of the nineteenth century, and the subject of some of the most artful political analysis in the literature of statecraft. But the basic assumption in this approach to geopolitics is that a country's actions are limited by a fixed and objective geography. Napoleon famously remarked that to know a country's geography was to know its foreign policy; that was certainly true in his time, and it still holds in many cases. Germany's location on the European plain, for instance, provided it with no natural barriers to stop attacks from the Warsaw Pact countries during the cold war.

But the flaw in this conception of geopolitics is that it generalizes across all epochs. It leads to conclusions that, however important, are not universal: sea power is superior to land power; whoever controls the heartland of Asia will control the world; Taiwan is endowed with enduring strategic significance and must be defended at all costs. These generalizations presume that what happened in the past will happen in the future. The north European plain is no longer an invasion corridor. That is what NATO expansion was all about: pushing back Russian armies one thousand miles to the east and creating buffer zones—Poland, Belarus, and Ukraine—to change an invasion corridor into an invasion barrier.

We have been told so many times that the world is shrinking that it's useful to be reminded once in a while that it is still 25,000 miles around the equator, and 5,500 miles from San Francisco to Beijing. Instantaneous worldwide communication and nonstop jet travel have certainly revolutionized commerce, politics, and even warfare in many ways, but armies don't travel over the Internet. After the United States defeated Iraq on the other side of the world, the view that geography was irrelevant became especially widespread. Jean Baudrillard, a leading postmodern French philosopher, argued the case in a book called *The Gulf War Did Not Take Place*. For Baudrillard, the war was fought in cyberspace, not physical space. Its "fronts" were on the evening news as video game replays of American missiles surgically taking out Iraqi targets. The physical movement of troops and vehicles on the ground, in the air, and at sea was overtaken by the imagery of war on a two-dimensional screen. To U.S. military commanders, time and distance were major impediments, but seen on the screen these factors vanished. The ability to edit the video clips gave the war a unity and rhythm different from the one experienced in the battle zone itself. As edited, the war was made to fit the evening schedules of the television audience.

The view that time and distance no longer matter has had a subtle impact even on the thinking of professional military officers. The former deputy chairman of the Joint Chiefs of Staff, Admiral David Owen, has argued that "information dominance" is the key to future military success. The side with the ability to see everything on a battlefield, whether it moves at night or under camouflage, whether it is underground or underwater, would seem to have a decisive advantage. If information dominance had been a possibility a half-century ago, there would have been no Pearl Harbor, and also no D-Day.

The danger in this view is the notion that information can substitute for force. The Western fixation on technology pushes in this direction. In the 1930s strategists believed that the airplane would be the decisive weapon in all wars to come. In the 1950s atomic bombs supposedly made obsolete all previous military thinking. Today it is information dominance, in which the United States is said to hold a commanding lead because it has pioneered computers and the networks to link them. Soon military comparisons will enumerate a nation's computing power along with its tanks, ships, and airplanes. But such assessments tend to blur the difference between war games and war. The most powerful computer in the world is no substitute for a well-armed and well-trained army in the right place at the right time.

The "death of distance" argument was taken to another level by the futurists Alvin and Heidi Toffler in their book *War and Anti-War*. They imagined a new age of warfare in which one side's computers would attack its enemy's hard drives with viruses, logic bombs, and Trojan horses—crippling electric power grids, banks, and telephones. The constraints of time and distance that have faced every general since Hannibal would be magically overcome.

Cyberspace replaces physical space; war is made bloodless, to conform with modern Western sensibilities about casualties and guilt.

But this isn't a very useful way to think about warfare in the real world. Airplanes don't fly in cyberspace, and armies don't travel down the information superhighway. For Western powers to influence events in Asia directly, they have to get there first. It takes twenty days to sail an aircraft carrier from the eastern seaboard of the United States to the Persian Gulf, thirty-five days from the Pacific coast. If the carrier is vulnerable to missile attack or other threats after it arrives in the Gulf, the risks for the United States go up dramatically.

A better example of the "death of distance" came in 1998, when North Korea fired a missile across Japan that splashed down in the Pacific Ocean. This event ended Tokyo's separation from Asian military space. The missile shot turned the Japanese archipelago from a zone of sanctuary into a target zone, ending a half-century during which Japan could plan for its defense without taking into account its geographic location off the coast of Asia. No Japanese white paper could have accomplished this. Only the concrete reality of North Korea's missile could do it.

But it's not just the military balance that's changing. Indeed, the interesting consequences are not even military at all. They are political. Redrawing military space redraws the political map as well. Asia's difficult terrain and, until recently, its relatively primitive transportation and communications networks tended to soften and diffuse political changes. Powerful social forces were dissipated in its vast rural territories, leaving Western outsiders relatively untouched unless they unwisely chose to get involved.

Between 1959 and 1962 Mao Zedong's Great Leap Forward threw China into turmoil. This mad scheme to achieve overnight industrialization so upended the economy that it led directly to the death of 30 million Chinese by starvation. Yet this disaster had no effect outside of China. The year 1962 is recalled as a dangerous one, but not because of anything that happened in China. The outside world suffered the staggering deaths there with indifference. What made 1962 dangerous was the arrival in Cuba of Soviet ballistic missiles, which altered the military and political space of North America and nearly triggered a thermonuclear war between the superpowers. These missiles turned an annoyance, Cuba, into a humiliation and a clear and present danger for the United States. It would do well to recall the U.S. reaction to Soviet missiles in Cuba when trying to understand how Asian countries feel about the danger that their enemies could obliterate them with the push of a button. Missiles in Cuba could be fired at any time, and this made the danger continuous. There was no advance mobilization or preparation required to use them. As such, they were a constant psychological check on the freedom of the White House. That was what President Kennedy thought, and why he was determined to get rid of them. At the time some observers argued that, since the Soviet Union already possessed nuclear weapons and delivery systems on its home soil, these missiles didn't change the fundamental military balance. They were wrong, yet exactly the same arguments are made today about Iranian or Pakistani missiles.

In 1962 few outsiders worried about famine in China. Compare this to the world's reaction in 1997 and 1998 to starvation in North Korea. North Korea's leaders inflicted this disaster on their own people by their extremism in pursuit of an insane economic program. But this disaster is being noticed in the outside world. The world is no more humane than it was thirty-five years ago, and also no less realistic. But the United States has a nightmare: that North Korea's internal disaster will explode over Northeast Asia. To prevent that, Washington must sustain a repugnant regime. Food and energy donations now prop up the most repressive government in the world because North Korea's missiles and chemical bombs can reach South Korea and Japan. If North Korea had the limited military reach it had in 1975, it is difficult to believe that government leaders in the United States and Japan would be rushing to its assistance. Rather, it would be back in the queue with African nations lining up for food aid from the international relief agencies.

In the more compact Asia of the year 2000, where commanders control missiles rather than peasant divisions, social eruptions have echoes beyond national boundaries. Until the nineteenth century, European wars and social revolutions had little effect outside Europe because the primitive military organizations in those countries couldn't project power very far. When the Thirty Years War of the seventeenth century devastated Europe, China and the Islamic world were left untouched. But after the Industrial Revolution, Europe's internal dislocations were amplified. With armies, navies, territories, and colonies on all continents, Europe's two great wars of the twentieth century became world wars. Asia is now going through a comparable transformation.

MAP MAKERS AND MAP TAKERS

Continents have an important place in strategy and in politics. They scale military forces, as in the "intercontinental" ballistic missile (ICBM) and bomber. The European Union and the North American Free Trade Agreement are important institutions built around continents. A "Eurasian Union," one that incorporated both Denmark and Cambodia, would be inconceivable. And just as each era rewrites its history, so it redraws its maps.

The implications are far more significant than a change of government or policies. The cold war radically changed the perceived spatial relationships among three continents, North America, Europe, and Asia. This reconception in turn required new designs for military forces. Nuclear-armed missiles compressed space and time, bringing the United States into proximity with the Soviet Union. Before the cold war, no one thought of these two countries as being near each other. From almost anywhere in the United States, Russia seemed almost inconceivably remote. Viewed from across the Atlantic and Pacific Oceans, Europe and Japan were far away, and Russia was beyond them. But that's not the path that bombers and missiles take. A B-52 taking off from North Dakota would have to fly only 5,000 miles to reach targets in the Soviet Union. An ICBM launched in Siberia on a polar trajectory would be over American territory in thirty minutes. Looked at over the North Pole, the Soviet Union and the United States are neighbors. During the cold war maps based on polar projections appeared in every newspaper and atlas, and they all clearly showed how near the Soviet Union was to North America. Suddenly this became the correct way to see the world. Accepting this new projection became a test of intellectual seriousness. Ridicule was heaped on the Mercator projection. Mercator maps were adequate for oceanic navigation because their distortions around the equator were small. But the cold war recentered world military geography from the equator to the North Pole. The Mercator map also gave true bearing for navigation. But that was for the old technology of the ship. The intercontinental bomber and missile required a new map.

The new proximity of North America to the Soviet Union led Americans to a new geography. Elementary schools deep in the heartland, surrounded by nothing but America for thousands of miles, began holding air-raid drills. In the 1950s every American school child was taught that Greenland was not really bigger than Australia, that the Mercator map lied, as if a sixteenth-century cartographer was at fault for not anticipating the cold war. With its distortion of northern regions, Mercator projections obscured relationships that military technology made it imperative to understand. Alaska was a remote hinterland on the Mercator map, but on the polar projection it was a cold war front, and soon enough it became a state. Great-circle routes of airplane travel were taught in the schools. And the U.S. military started a massive redeployment from its natural southern homeland northward to Maine and North Dakota, to Canada, Alaska, Iceland, and Greenland. Air bases there could get a jump on the Soviets in a nuclear war.

The Distant Early-Warning (DEW) and Pinetree Lines were radar networks strung like Christmas-tree lights across Canada. They replaced coastal defense batteries and western forts as U.S. frontier defenses, becoming part of the American lexicon, metaphors for alertness and foresight. Marshall McLuhan in the 1960s dubbed his futuristic media newsletter the *DEW Line Report,* suggesting an early-warning radar of social change. The submarine *Nautilus,* whose very name was borrowed from Jules Verne's futuristic story about the technological exploration of a hostile environment, sailed under the North Pole icecap, continuing, to vast national acclaim, the pioneering tradition begun by Lewis and Clark.

The cold war also changed the geography of Europe, which suddenly seemed small and vulnerable compared to the United States and the Soviet Union. Books with titles like *Europe Between the Superpowers* demoted it to a theater in the larger global competition between Washington and Moscow. Only the weakest expressions of Europe's own interests were possible in this zone. Yet just ten years earlier Europe had deployed the most powerful military forces in the world. In 1940 Europe had been the initiator of terrible events, and other parts of the world had been made unwilling extensions of its quarrels. By the 1950s Europe no longer had the power to initiate events on the world stage. It went from being a map maker to a map taker.

Asia has seen many changes in geography as well. But only recently has Asia been its own map maker. Asia was created, named, and bounded by Europeans. The lands it comprises had no common identity until one was constructed by Europeans in the processes of colonization and domination.

Continents were originally conceived of as large continuous masses of land. At first, there were only two continents, the Old World of Europe, Asia, and Africa and the New World of the two Americas. In the eighteenth century, however, there began a more systematic attempt to make the world fit Europe's conceptions of order. More continents were added, with definite borders replacing the vague boundaries of earlier times.

In most cases the dividing lines were clear. Europe was separated from Africa by the Mediterranean Sea. The Atlantic Ocean lay between Europe and the Americas. Africa was separated from Asia by the Red Sea.

But the problem was Asia. No natural geographic feature such as an ocean existed to show where Europe ended and Asia began. Topographically Europe and Asia could have been constituted as a single entity, one continent, what today would be called Eurasia. But this isn't what happened. Map makers severed Asia from Europe at the Ural Mountains.

Modern geographers have argued that this definition of continents is simplistic and misconceived. But European map makers knew exactly what they were doing: drawing their maps to fit the political climate of their time. Europe the map maker redrew Asian space in a way that abetted its larger strategic purpose by erecting a barrier to separate itself from Asia. In so doing, it raised the question of whether Russia should be categorized as European or Asian, a subject with potent strategic implications.

RUSSIA IN ASIA

The existence of Asia made Russia into a European country. The Ural Mountains mark lands that are identified as different and, for eighteenth- and nineteenth-century Europeans, weaker and backward. This separation bolstered the Russian sense of its European identity. It was no accident that the partition along the Urals came about only after Peter the Great began his campaign to modernize Russia and make it part of Europe.

Russia at this time was taking over backward Asian kingdoms just emerging from centuries of Mongol rule. In the eighteenth century, Russia rapidly colonized Siberia, then Central Asia and the Caucasus. Petropavlovsk, on the Kamchatka Peninsula, was founded in 1752; by 1790 all of Siberia was under Moscow's control. The major European powers, France and Prussia, resisted Russia's goal of western expansion. But in the East, Russia's superior military organization quickly conquered and scattered the soldiers of the loose-knit tribes.

Not wishing to be tainted in Western eyes by these backward regions, but very much wanting the riches they offered, Moscow in effect created two Russias. One was modern, dynamic, and European; the other was Asiatic, barbaric, and ripe for the civilizing effects of Russian rule. The Russian heartland centered on Moscow became more European by relegating its Asian acquisitions to an entirely different geographic space. It created an exploitation and extraction zone and a notorious penal colony whose imprint lasts down to the present.

An accident of timing helped Russia's expansion. Eighteenth-century Japan, deep in self-imposed isolation from the rest of the world, was no barrier to Russian colonization in Asia. Had Japan allied itself with one of the Western maritime powers and cast off its technological shackles a little earlier, there can be little doubt that the Russian advance to the Pacific would have met much greater resistance. It is likely that the entire face of Asia, and the world, would look quite different today if Russia had been kept out of Asia.

The Russian-Asian divide even worked its way into American atomic war plans. In the 1980s U.S. leaders were concerned that Moscow might not care about losing millions of people in a nuclear exchange if they were Uzbekis, Kazakhs, or other non-Russians. Fearing that the loss of Tashkent in a nuclear war might not count for much in the Russian calculus, Washington aimed more of its missiles at the European heartland, to guarantee destruction of what was thought to be held most dear.

In a strange way the attempted nuclear amputation of Russia from Asia actually played out in peaceful form during the breakup of the Soviet Union. When the Soviet Union fell apart in 1991, Kazakhstan and the other Central Asian republics were cut loose. Russia lost its empire and with it its place in the world. Where once Russia was modern and advanced—meaning, essentially, European—now it is looked down on by the Japanese and the Chinese. This view of Russia has actually been evolving since the 1970s, the result of its great failing in the second half of the twentieth century, its inability to become a Pacific power. Because Russia conceived of its Asian territory as a region to exploit rather than as a gateway to the Pacific, the economic rise of the Pacific Basin bypassed the Soviet Union. The United States learned from the Asian business challenge. It restructured its economy and copied many Asian business practices. The Soviet Union learned nothing. The Asian economic boom that was apparent to the rest of the world by the 1970s was entirely overlooked by the Soviet Union.

A critical question is what the current Russian chaos means for the Asian part of the country. If economic chaos and political disorganization continue, China, South Korea, and Japan could pull the Russian Far East into their orbit, breaking up the Russian state yet again. For almost the entire twentieth century the Russian Far East has looked to Moscow, but those ties are fraying. Almost no one today envisions continental Northeast Asia as a disputed territory. But only 8 million Russians inhabit the territory from Lake Baikal to the Pacific Ocean; this area, almost as large as the continental United States, is a vast empty space next to China, with its 1.2 billion people. Something like half a million Chinese move into Russian areas each year already, following the collapse of Soviet border controls in the early 1990s.

Holding this area together is not as simple as it was under the czars, much less Stalin. The Russian state is in chaos. While Westerners are focusing on Moscow, disorder and discontent grow with distance from the capital. The Russian army shrank from around 4 million at the time of the breakup in 1991 to 1 million in 1998, and its capabilities probably declined even more precipitously. Officers and enlistees are unpaid, food is short, and housing is squalid. The distances this pathetic army must defend are vast. The distance between Fort Bragg, North Carolina, and Kosovo is about 4,500 miles. This is as far as the distance from Moscow to Kamchatka.

It might not take an outright revolution for the Russian Far East to spin off into independent states, becoming a "Ural Republic" or an "East Siberia" aligned with China, Japan, or South Korea. There are many ways for a state to break up. In 1989 Russian power collapsed in Europe without open warfare. In a similar way, the Russian Far East could separate from Russia gradually as the Russian army loses its ability to hold the state together.

EUROPE IN ASIA

While the Russians colonized Asia by land, Europe—blocked by the powerful and implacable Ottoman Empire—did it by sea. In the process, Europeans invented a new technology, the heavy seagoing warship, and a new type of geopolitical structure, the overseas maritime empire. The most important impact of Western military technology and organization on Asia was not in the battles it won but in how it changed the mental geography of the continent. Distance took on a new meaning when it was spanned by steamships and railroads rather than by sailing junks and caravans. The Suez Canal is a good illustration. Before it opened in 1869, Europe's heavy armies were of limited use in Asia. They could not march overland through the Ottoman Empire—enfeebled though it was by the midnineteenth century compared to its earlier glory days—because of the harsh desert terrain. They needed enormous logistical support to travel any significant distance—ammunition, food, water, and shelter. Given the technology of European armies of the time, this was impossible. But with sea access, and the canal, a path was driven right through the empire. A new front was created in what had once been a rear area, the Red and Arabian seas. What was distant, India and China, became closer.

European military technology certainly helped win battles, but its more important effect was to reorganize geopolitical space. The sea lanes leading to and around Asia became the foundations of European military power in Asia. The context of relations changed from cooperation to coercion. A fragmented series of British colonies and bases, strung from South Africa to India to Malaya and Hong Kong, turned into an empire. Steamships traveled faster than sailing ships, obviously, but even more important, a loose collection of ships and ports became a unified network capable of mutual defense because the schedules of ships no longer depended on the wind.

The undersea submarine telegraph wrought a similar transformation. For the first time in history, a naval force enjoyed instantaneous communication (while in port). Previously, military orders had to be either carried aboard ship or sent by couriers overland to ports for the fleet. The effect of steam and the telegraph was to create a taut sensory network that allowed Britain to disperse its navy over a vast expanse yet respond to local crises in a timely way. When the fleet was needed at any particular trouble spot, it could steam there quickly and reliably. Dispersed ships could be based forward, distant from the mother country, and ordered by telegraph to mass for local superiority. Thus could a tiny force control a vast area.

The key geopolitical structure here was the network. Network timetables that listed what the British navy could reach, with how many ships, in a given amount of time, became the key to controlling Britain's vast empire. Before the telegraph and the steamship, such timetables were impossible.

Britain had no monopoly on these technologies. Together with another revolutionary technology, radio, these developments, were to reorganize East Asian geography once again. Early in the twentieth century Japan began a major program of constructing warships. This so alarmed the West that it responded with an early international arms control agreement, the Washington Naval Agreement. The agreement was not very successful, in part because it failed to take into account the development of the aircraft carrier and the way in which Japan would integrate its land and naval forces.

The aircraft carrier allowed a country to move its airpower forward without land bases. More than anything else, it allowed surprise. The aircraft carrier allowed the air base to appear suddenly, seemingly out of nowhere, as the success of the Japanese at Pearl Harbor demonstrated. Five Japanese aircraft carriers sailed north of the usual Pacific sea lanes, staying out of sight and operating under radio silence to avoid detection. They turned south for the attack, catching the unwary American defenders in Hawaii by surprise.

The role of the integration of the Japanese navy and army in transforming Asian military space is much less appreciated. Many historical accounts describe the feuding and bureaucratic warfare between the two services. But there was enough integration to allow Japan to overrun East Asia in a matter of weeks in late 1941 and early 1942. On a map, Tokyo looks to be 3,000 miles from Singapore. But the distances separating the intervening islands and coastal ports in China and Vietnam were small. The Japanese navy had to haul the army only relatively short distances to leapfrog all the way to Singapore, where the British were completely routed. No military force in history had put sea and land forces to such a use, and the effect on the British and Americans was devastating.

For the United States, Pearl Harbor revolutionized its concept of Asian geography, literally overnight. While there were many before 1941 who saw the need to protect American interests by having a military that could shape events in East Asia, it was not until Pearl Harbor that this need finally sunk into public consciousness. The effect was to alter every aspect of American national strategy. Most tellingly, the vocabulary of defense changed overnight. Before World War II, *hemispheric defense* had been the overarching military framework of the Republic since its founding. After Pearl Harbor, a new term, *national security,* came into use. It covered a much broader range of possible threats than direct attacks on the American territory itself. The new term recognized that dangers to allies, or even a buildup of hostile military power in distant parts of the world, could affect the United States in profound ways. It was an expanded spatial conception of defense that has endured to the present time.

ASIA IN THE COLD WAR

The cold war had its own geography. For the United States after World War II, Asia was not some outer region beyond Europe. The very term *Asia* was itself too big to fit into the spatial requirements of the cold war, which had to do with containment, not colonization. The new map of Asia was built on subregions of the Asian rim, acknowledging that inner Asia was necessarily controlled by the adversary.

This view was given reality by government and foundation support for "regional studies" at universities. Academic programs flourished around the culture, society, and politics of regions relevant for cold war strategy. *Asia* disappeared in the mental map of the West; in its place arose the Middle East, South Asia, Southeast Asia, and Northeast Asia. For example, before the 1940s Thailand, British Malaya, and French Indo-China were distinct entities, but grouping them under the heading of "Southeast Asia" conveniently wiped out their colonial heritage, removing a memory that worked against cold war containment policies and replacing it with a neutral designation that suggested common interests and fates. This designation presaged the "domino theory," which held that if one country in Southeast Asia fell to communism, so would all of the others, as one domino in a chain knocks down the others. Likewise, the Middle East, a term originally coined by the American naval strategist Alfred Mahan in 1902 to describe the lands surrounding the Persian Gulf, also received a new geographic unity in the cold war. The Eisenhower Doctrine of the 1950s stated that the United States would not allow a country in this area to go Communist. Geographic designation drove strategic declarations.

This was not the purely psychological change it is often made out to be. It may have started out that way in the 1940s, but it was surprising how soon the new geographic designations shaped the military dynamics of the cold war, the movement of forces, the location of bases, and the thresholds for military action. This geographic regionalization even affected American nuclear war plans. This is one of the most neglected fields of international politics: the way in which geographic designations create political boundaries that, if crossed, can lead to international tensions, even war.

In the cold war crucial subregions were given military command structures to ensure their defense. For Europe, the western appendage to Asia, there was the North Atlantic Treaty Organization (NATO). In the 1950s NATO was cloned and applied to the Middle East (the Central Treaty Organization, or CENTO) and Southeast Asia (Southeast Asia Treaty Organization, or SEATO). In Northeast Asia separate treaties linked the United States with Japan and with South Korea, two countries whose history of animosity earlier in the century precluded grouping them in a mutual defense pact.

Cold war escalation was shaped by this geography. Escalation confined within a subregion was one thing, but its potential spread to other regions was considered very serious. Such "compound" or "horizontal" escalation played on the fundamental weakness of the United States, which faced the much larger land armies of the Soviet Union and China. A two-theater war would have required the American army to split its forces, defying military wisdom at least as old as Hannibal.

This situation was recognized as a critical threshold for the transition to nuclear war, and it was the reason the United States built tens of thousands of tactical nuclear weapons. If Moscow had simultaneously attacked Western Europe and the Middle East, it was almost certain that nuclear weapons would have been fired at the advancing Soviet armies. American war plans called for waging a conventional defense in any one subregion. But if escalation spread to other subregions, nuclear war was almost ensured. Given that Europe was the most important of the cold war subregions, this rule implied early nuclear strikes against Communist attacks in Asia. Otherwise, the army in Europe would have to be swung to Asia, leaving NATO unprotected.

Like the British earlier in the century, the United States built a geopolitical structure that was much more than a collection of military units and bases. It was a sensory network with rules and interrelationships among its parts, extending even to nuclear forces. Understanding how new military technologies give rise to geopolitical structures such as NATO is critical to understanding the slow changes that concern us here.

THE CREATION OF THE *PACIFIC BASIN*

Over a hundred-year period the concept of *Asia* had been created by Western maritime power, and then deconstructed by the cold war. Asian space again changed in the 1970s as a general economic boom spread over Asia's rimland nations. Thus began a trend back to a larger geographic entity, the *Pacific Basin*.

The regional mindset imposed by the cold war overlooked this critical development. High rates of growth first appeared in Japan, and then in South Korea. In the 1980s the economic boom spread to Thailand, Malaysia, Indonesia, Singapore, and eventually China. U.S. trans-Pacific trade grew larger than trans-Atlantic trade. The geopolitical theorist Herman Kahn of the Hudson Institute argued that the Pacific Ocean was becoming a great connector, rather than a great divider, of nations.

The common thread of economic growth among nations that had little else in common gave rise to this new geopolitical structure, the Pacific Basin. It stood for economic dynamism, market capitalism, and a national emphasis on business. Over time the Pacific Basin has grown to include more and more states; in the 1990s a new term, the *Asia-Pacific Region,* extended the concept to India.

The Pacific Basin is a zone where, in theory, threats of force are unthinkable. Military threats would only undermine the region's claim to unity, its economic dynamism. A geography of finance and commerce has no place in it for war.

There is one notable exception. One great power must protect this fabulous success story. Nothing must be allowed to reverse this great transformation, neither local conflicts that get out of control nor an emerging power trying to set things on a different course. Only the United States can play this role. The Pacific Basin is the ideal military geography for the United States because it assumes a compliant acceptance of American military dominance in Asia. Those who contest this, like Iraq, are isolated by the United States and its Asian allies. This, at least, has been the theory in the 1990s. Thus continues a centuries-long tradition of the greatest military power in Asia not being Asian at all. How long can this tradition last?

LEARNING FROM GENGHIS

Asia stretches military forces. The great distances and the varied topography challenge military organizations to overcome both. The greatest military power in Asia, the United States, is automatically affected by this challenge. As Asia stretches its military, those forces spill out into spaces that once were solely utilized by the American military. As missile ranges increase, as navies move farther from shore, and as military satellites track targets, increased interaction and friction are inevitable and unavoidable if both parties occupy the same space.

Asia's tendency to stretch military forces is not new. The Mongols conquered all of Asia, including China and Persia, and penetrated to the heart of Europe. Theirs was the largest land empire in history, put together in only a few decades of the thirteenth century. It was based on horsemen using speed and stealth to outflank opponents, fighting in self-contained formations but directed by the Great Khan. It is said that daily situation reports were sent back to his court in Mongolia three thousand miles away, and that even at this distance he had a better picture of local conditions than did his opponents. To wage war in Asia requires forces that can be effective over its vast distances.

Traditional European armies have proved useless in Asia; their place has been usurped by maritime forces built around navies and light infantry. The United States defeated Japan in World War II with a new combat organization consisting of amphibious forces, submarines, and aircraft carriers. The tanks and heavily mechanized units that were used in Europe played only a small role in the Pacific. Three-fourths of the U.S. army's eighty-nine World War II divisions were put in Europe. Nineteen Marine divisions carried the brunt of the fight against Japan. The pattern begun by the Mongols and the European seapowers prevails today. Outside powers disperse their forces in Asia, relying on communications and mobility to mass them as needed. The U.S. military presence in Asia is mostly on and under the water, spread relatively thinly across 12,000 miles, with bases in the Middle East, Diego Garcia, Guam, South Korea, Japan, Hawaii, and Alaska. In NATO Europe, in contrast, American forces have been much heavier and more compact.

Asia's immense size has other consequences. For most of history alliances never mattered much in Asia. It was all but impossible to rely on allies for support. Distances and terrain made it impractical to send military assistance in time to make any difference. This is why there was never a NATO in Asia, or a Warsaw Pact, and why even the NATO clones such as CENTO and SEATO never worked there. They could not overcome the long historical tradition of nations fending for themselves.

Following the pattern that dates back at least to the Mongols, many Asian countries are now expanding the reach of their armed forces. They are building missiles and weapons of mass destruction, not infantry forces. China, for example, has been gradually moving forces outward toward the Spratly Islands in the South China Sea, using naval patrols and marine deployments to attain virtual sovereignty there. The Spratly Islands, 800 miles southeast of China, are usually looked at for their potential to hold undersea oil deposits. They are disputed territory claimed by Vietnam and the Philippines as well as China. Whether there is in fact oil on the seabed around them is something only time will tell. But looked at more broadly, an interesting pattern is revealed. In its effort to secure the islands, the Chinese navy is stretching its abilities to operate far from the Chinese coast. Air combat missions are coordinated with the surface fleet. Refueling at sea is practiced. Weather satellites are tied into fleet operations. Communications and intelligence, to warn of U.S. Navy ships approaching the islands, are perfected. Whatever happens with respect to oil, the more enduring consequence of the Spratly Islands dispute will be to increase Chinese know-how in projecting force over long distances. A fleet that can be sent to the southeast can also be sent to the northeast. A navy that can operate 1,000 miles from home is one that can be stretched to operate even farther from shore.

In the past only outside Western powers had sensory networks in Asia, communications grids using satellites and underwater listening devices to find out what was going on. Now China, India, and Israel are all beginning to build such networks, which, although primitive compared to U.S. communications, mark a profound technological turning point. Japan recently announced that it is launching four intelligence satellites of its own, to more carefully watch events in East Asia without the filter provided for the intelligence it shares with the United States. China in particular is getting better communications, with its far-flung military deployed on borders with Vietnam, India, Kazakhstan, and Russia, as well as in the South China Sea. China even has an intelligence station in the Indian Ocean to watch the Indian navy. All of these dispersed units must be coordinated and linked to centralized military headquarters.

Indian armed forces are also being stretched. Indian military geography for centuries was shaped by invaders coming across the northwest frontier and advancing into the richest part of the country. The Himalayas secured the northeastern frontier from Chinese attack. But in the northwest the large Punjab plain invited attack by land armies. This plain is in the western part of a large horse-shoe-shaped agricultural belt extending from Calcutta all the way to Karachi. The area is so flat that from Calcutta to Lahore, a distance of 1,300 miles, the elevation changes by only 1,000 feet. As has happened time after time in India, cavalry forces coming from the Khyber Pass or other portals can travel by way of Delhi to the Bengali capital of Calcutta without meeting a single natural barrier. A network of forts has been in place since time immemorial to block this traditional gateway, beginning with Peshawar and Rawalpindi (both in present-day Pakistan). Given this vulnerability, there was nothing for an Indian navy to do, and it was always deemphasized in favor of the army.

The northwest still presents a vulnerable invasion corridor, and Delhi has focused on Pakistan since both states were created in 1947. Now, however, an entirely new geographic expansion is stretching the Indian military to greater distances, and in new directions. India now watches China, once a country largely irrelevant to its security because the two were separated by the highest mountains in the world. Until the early 1990s India relied on Moscow for satellite pictures of Chinese troop dispositions in Tibet. These came in every few months and put India at the mercy of Soviet intelligence. But now Indian satellites and reconnaissance aircraft provide this information directly.

The changing focus of the Indian navy pulls the nation's military in yet another direction, to the east. The navy has traditionally had three command centers, all on the mainland. Recently a new Far Eastern Naval Command has been set up in the Andaman Islands, 750 miles east of the mainland, to counter a Chinese naval presence in the area. To keep track of the Chinese forces, India has installed extensive surveillance and intelligence monitoring, stretching its forces in the opposite direction from their traditional focus on Pakistan and placing enormous new demands on the Indian navy to communicate and coordinate in an area where only a few years ago the Indians were, for all practical purposes, blind. It is highly likely that this dispersal will require Delhi to reshape its military command and control system, just as Beijing has done.

Organizational stretch drives the strategy of many Asian nations as they seek to monitor potential trouble spots beyond their national borders and to develop the ability to reach those places with military force. This factor has mostly been overlooked, however, in the West, where analysts tend to view strategy in terms of a grand long-range political plan to spread influence. Speeches of national leaders are carefully examined for clues about their future intentions. But it is the larger structural change that better explains why Asian countries are rushing to develop increased military and communications capabilities. Increasingly, an all-azimuth concern for security is forcing countries from Israel to North Korea to confront and overcome geographic obstacles, especially distance. The Indian navy is covering the eastern Indian Ocean, where ten years ago only the Americans could operate. China is building a communications link to its armies in the far west, and to its forces in the South China Sea. Israel uses satellites to watch Iraq, Iran, and Pakistan.

THE COMING CRISIS OF "ROOM" ON
THE EURASIAN CHESSBOARD

The Pacific Basin constitutes a geography of one superpower and many subordinate states voluntarily concentrating on economic, not military, development. But it is falling apart as Asian industrial power becomes aligned with Asian military power.

The United States has not recognized the significance of this development but has tried to preserve the geopolitical structure of the past fifty years, decoupling Asian industrial capacity from military force. Washington sees its interest in Asia as preventing the rise of any power or group of powers that could challenge its leadership. Its rhetoric is meant to suggest that the growth of any outside military power is a challenge to a basic order, the Pacific Basin, that has brought prosperity to all who have chosen to participate in it. In concrete terms, this policy translates into holding on to bases, stopping the spread of missiles and weapons of mass destruction, isolating renegades, and managing pivotal states like China.

But to play this game on the Eurasian chessboard is to fail to appreciate how much has changed since the end of the cold war. The power of many of the pieces is rapidly growing through the acquisition of weapons of mass destruction. The chessboard is shrinking as pawns take on the range of knights and bishops. Previously unrelated parts of the board are now highly linked, making it hard for a power whose forces are dispersed over an enormous area to achieve local advantage. For example, the United States nearly always looks at China as an East Asian country. But China's supply of missile parts to Iran greatly increases the pressure on U.S. forces in the Middle East. Whether China consciously intends to do so or not, Chinese sales of military technologies to Iran worsens the U.S. strategic position in Asia. Building up defenses against Iran takes resources away from East Asia, unless the overall U.S. defense effort is increased. In short, playing the game without recognizing that the chessboard has changed can only lead to disaster.

This does not mean that the United States should withdraw from the game. To retreat from Asia—or to disarm to the level of France or Britain, with forces adequate only for short-distance peacekeeping—would invite arms buildups throughout the continent. Countries would feel threatened, and this military pressure would force some in the Middle East, like Saudi Arabia, to succumb to blackmail from Iran or Iraq. South Korea and Japan would very likely go nuclear. Pakistan would be in an even more desperate and trigger-happy condition than it already is. And Iraq, Iran, and Syria would combine their missile power against Israel.

The challenge for the United States is to play the game on a new chessboard. Indeed, recognizing that the game is taking a more dangerous turn creates new opportunities. The old techniques of power—fixed bases, arms control, and "managing" China—have played out their usefulness. They can be kept for a while, but they cannot be the foundation for future American military presence in Asia. The value of fixed bases is certain to decline. No new arms control regime for weapons of mass destruction is likely to be anywhere near as successful as the one now ending. And there is no new China policy waiting to be discovered that will move China onto the track that Americans would ideally like to see that country take.

The shrinking Eurasian chessboard also creates a structural crisis not seen before. Simply stated, Asia is running out of room: room for political and military maneuvering; room to buffer relations between armed powers; and room to insulate domestic crises from spilling over into the international system. This is a critical point: the crisis of room has major implications for how military forces are structured and operated, and for how basic strategies and agreements are formed.

Eurasia's size in the past allowed one or two countries to enlarge their military without any undue danger. In the nineteenth century Britain and Russia met in Afghanistan and Central Asia in what was called "the Great Game." In the twentieth century the United States confronted the Soviet Union. In both cold wars each side acted as if such confrontation were the axial event in its foreign policy. Meticulous plans were built for the clashes between giants.

Britain had plans for using 100,000 camels to supply a war against Russia in Afghanistan from India. The United States had plans for using tens of thousands of nuclear bombs. Neither plan was ever executed. The parties didn't want a war so incredibly expensive and destructive. Both sets of antagonists learned to live and let live, with informal agreements that clearly spelled out how far each could go.

It is useful to keep this in mind when a forecast is made of a coming war between China and the United States. The basis of this pessimistic forecast is a clash for domination of Asia. China at the moment shows none of the aggression toward the United States that the Soviet Union did. A major military confrontation with Washington would be highly upsetting to Beijing's economic and other interests.

But more important, concentrating on a fundamental clash of interests between two powers misses the more important structural transformation in Asia, the expansion of military reach in an area whose stability used to depend precisely on the inability of countries to attack each other's heartlands. Technology's relationship to geography is shrinking everyone's maneuvering room, political and military. Tensions increase as once-isolated domestic pressures spill across national borders.

The spread of the bomb is unfortunate. But it is tolerable if it is confined to one or two countries, and if they do not use it for anything more than symbolic purposes. When eight or nine contiguous countries get the bomb or missile-mounted weapons of mass destruction, the situation changes. The chessboard shrinks, just as it shrank in Europe at the beginning of the nineteenth century. Diplomats then failed to appreciate that industrialization, applied to war, had eliminated maneuver room. It made the consequences of any conflict far worse than anyone had imagined. Yet diplomats blithely went along as before, playing the old game on a radically constricted chessboard.

A crisis of room is a possibility that was noted many years ago by the great Hungarian-American mathematician John von Neumann. Von Neumann was undoubtedly one of the greatest minds of the twentieth century. He invented the mathematical theory of games and helped design the hydrogen bomb. His design for memory storage devices was a breakthrough in early computing. Writing in 1954 for *Fortune* magazine, von Neumann argued that the space in which technological progress took place was becoming undersized and underorganized, not as a result of bad policies but as an inherent consequence of technology's relationship to geography.

Von Neumann's contention was that geographical and political space provided a kind of safety mechanism for technological advance. Asia's agricultural society has had just this dampening effect on social shocks and political dislocations. The Great Leap Forward, monsoon disasters, and even the Vietnam War had little effect on the outside world. The basic structure of the Asian system went unchanged, and Asia's impact on the world was not that great.

The problem identified by von Neumann in 1954 was that political space was being outrun by the expanding influence of industrialization and military technology. This is an easy change to miss because it is gradual, and because of the way most people look at weapons. New weapons are usually seen as mere incremental improvements over earlier ones, or as changes in the military balance, the bean count of who has what. The effect of new weapons on the structure of international politics is easily overlooked. The

Gulf War showed this clearly. Before 1991 no one imagined that a country on the outer fringes of Israel's security space would attack it with missiles. In earlier Middle Eastern conflicts, in 1967 and 1973, Iraq had sent only a token ground force against Israel. But in 1991 Israel was struck directly by this fringe country. Suddenly the fate of the Gulf region became inextricably bound up with events in the Mediterranean. Had Iraq had a few more years to develop its mass destruction weapons, the very survival of Israel would have been at risk.

Von Neumann worried that the finite size of the earth would itself be a factor in increasing instability as more nations industrialized and expanded the geographic scale of their armed forces. In a cold war world with two superpowers, all kinds of agreements—including informal understandings about the geographic thresholds for nuclear escalation and self-imposed limits on new weapons—worked as well as they did because cold war geography provided maneuver room. Moscow backed down in Cuba, but heated up an insurgency in Vietnam. In this way, the superpower confrontation was shifted about from one place to another, all within self-imposed constraints on how far any one crisis could be pushed. The United States and the Soviet Union were only a few miles apart across the Bering Strait, but almost all of their confrontations took place in spaces with plenty of room.

But this is not the case when it comes to North and South Korea, Pakistan, India, Israel, Iran, China, or Taiwan. For them, missiles and weapons of mass destruction remove the buffer room they once had. Immediate dangers and opportunities on their borders are close enough to present real problems and are within the grasp of politicians needing an issue to mobilize supporters. When Iran fired a missile with the range to hit Israel, it did more than test a weapon. It shrank the Middle Eastern corner of the Asian chessboard even more than in 1991 and brought two very different political systems into proximate contact. Reducing the psychological distance between the two changed Middle Eastern politics forever.

The crisis of room on the Eurasian chessboard fundamentally changes the game for the United States. U.S. strategy since the cold war has focused on conventional regional war. The geographic size of this war can even be specified because it comes from the Gulf War and the Korean theater. A box with 200-mile sides nicely covers both of these regions. A 40,000-square-mile box defines the limits of current strategic thinking as well. Within it, the United States prepares for a conventional war in which nuclear, chemical, and biological weapons, and ballistic missiles are not used.

This plays to American strengths. No nation could stand up to the United States in such a war. So it is natural that the United States should do whatever it takes to shape the strategic environment this way. If conflict can be kept conventional and confined to this box, if people can be made to think of war and deterrence in only these terms, the psychological advantage to the United States is tremendous. Arms control, technology embargoes, and diplomatic pressures to limit the acquisition of new weapons all make perfect sense.

But what is taking place is something quite different. By changing the geographic scale of their armed forces to operate in a bigger box, or boxes, and by diversifying from conventional to weapons of mass destruction, Asian nations—especially China, India, Iran, Iraq, Pakistan, and North Korea—are overcoming the U.S. psychological and military advantage. Much like a business going into new geographic and product areas, Asia is expanding its military scale to larger areas and diversifying into nuclear, chemical, and biological weapons.

There is also a new check on the United States. In the cold war it was constrained by the fear of escalation. That was the check

against launching air strikes against Soviet missiles in Cuba. It was
an even more powerful impediment in Vietnam, where the fear of
Chinese intervention made leaders in four administrations conduct that war with a level of restraint that virtually guaranteed failure. Since the end of the cold war no such fear of escalation has
been present to put such restraints on the United States, with the
unfortunate effect of divorcing political from military strategy. The
Gulf War was turned over to military commanders in a way not
seen since the Spanish-American War. This can't continue in the
future.

The prospect that China or Iran or any one of a number of others could accurately target either the United States or its bases in
Japan or Saudi Arabia must result in a full-scale reassessment of the
American military posture in Asia. The problem facing the United
States is not one that can be solved by reproducing better versions of
existing military equipment. That has been the approach of the last
few years, but as we see in the next chapter, it is one that does not
come to grips with the different nature of these new technologies.

2   DISRUPTIVE TECHNOLOGIES

The country with the best technology does not always win, either in battle or in the struggle to impose a world order. North Vietnam fought with primitive technology and beat the United States. France had more and better tanks than Nazi Germany but was overrun in a matter of weeks. Britain had the finest naval technology in the world early in the century. Nonetheless, its empire fell.

The point can be made more strongly. Putting faith in a technology lead is often a fatal distraction, not, as one would expect, because morale and strategy are more important than technology. Nor is it that Western military technology has become too complex to use.

The danger lies in the failure to distinguish between technologies that help sustain leadership and those that undermine it by disrupting the status quo. This distinction is basic in all fields of competition. Microsoft dethroned IBM in the early 1990s not by selling better computers but by competing with a different technology, software, thus upsetting the market in which IBM's know-how had been dominant. Microsoft had a better software than IBM and pushed this advantage to the limit. Customers sought out the better software because it enabled them to do what they wanted on their

PCs, and pretty soon it was the software rather than the computer that developed into a brand name. IBM continued to sell good computers, but its competence in that area was rendered irrelevant by Microsoft's software; soon the PC became a commodity.

Disruptive technology changes the game. By upsetting existing advantages, it nurtures new skills and fosters different strategies. The resulting uncertainty shakes up the established order and changes the standards by which leadership is measured. Leaders whose dominance is threatened often do not recognize major environmental changes until it is too late. Using old standards, they assess their situation and conclude that they are safely ahead. But that advantage is measured by the standards that don't count anymore. Like IBM, such leaders stick with an outmoded paradigm. The ballistic missiles and atomic, chemical, and biological weapons coming to Asia are all disruptive technologies. They nullify Western advantages in conventional weapons. They restrict Western military access to Asia. They foster new talents—for example, brinkmanship using weapons of mass destruction—that undermine the West's conventional superiority. In short, they trump the Western technological lead.

The West can keep its technical lead over Asia. Its cruise missiles can strike with pinpoint accuracy. The West can have better tanks, bombs, and airplanes, measured in any head-to-head comparison. But disruptive technologies diminish the importance of these comparisons because they shift competition to areas where geographic and political factors work against the West.

THE DAWN OF POSTMODERN WARFARE

Biological weapons are a good example of a disruptive technology. They have a radically different profile from traditional conventional weapons, and this makes them hard to categorize in Western strategic thinking. Iraq's program is a case in point. The program was cut short by Baghdad's defeat in the Gulf War. But for this accident of timing, Iraq would have succeeded in building a germ weapon arsenal that could have attacked U.S. bases in the Middle East, rendering them unable to launch the kind of strike that decided the Gulf War. Iraq also could have killed a significant part of the population of Israel.

Iraq's program started in 1974, two years after it joined one hundred other countries in signing the Biological and Toxin Weapons Convention, a treaty that had no inspection, verification, or enforcement mechanisms. Throughout the 1970s, and during its war with Iran in the 1980s, Iraq undertook a very wide research program to find germ agents with field utility. Anthrax (a cattle and sheep disease that is fatal 90 percent of the time when transmitted to unprotected humans), aflatoxin (causing liver cancer), botulinum toxin (causing muscle paralysis), tricothecene mycotoxins (which induce high-speed projectile vomiting and nausea), wheat cover smut (an anti-crop agent to starve an enemy), and rotovirus (causing acute diarrhea) were some of the agents cooked up and tested in Iraqi germ laboratories. The stand-out agent here must be aflatoxin because it takes such a long time for the cancer to develop. The logic was that if the West attacked Iraq, it would retaliate so that even young people would get cancer years later in their middle age. Just before the Gulf War, Iraq started a crash effort to turn this research program into weapons that could be used to wipe out Israel. It is important to recognize that the program was mainly designed to kill unprotected soldiers and civilians in the rear, not frontline combat troops. In doing this, Iraq contemplated a different kind of war. MIG jets were retrofitted to fly by remote control, without pilots aboard. They were equipped with sprayers to drizzle anthrax over Israeli cities from 250-gallon storage tanks. Although such attacks were never conducted, one of the jets was actually test-flown.

After inspection by the United Nations, Iraq admitted to building 157 germ bombs for delivery from aircraft, by artillery and short-range rockets, and by pesticide sprayers. It also admitted to having 25 germ war missiles, modified SCUDs, of the kind that were fired at Israel and Saudi Arabia with conventional explosives. Of these 25 missiles, 16 used botulinum toxin, 5 used anthrax, and 4 used aflatoxin. These weapons were actually sent to the field at the start of the Gulf War, where they were prepared for firing on short notice. Amazingly, predelegated launch orders were given to the military commanders by Saddam Hussein. This meant that had certain conditions been met—an allied ground attack to end Saddam's regime—the commanders would have fired their germ-filled SCUDs at Israel without further orders.

Iraq's actual biowar stockpile was larger than its admission to UN inspectors. Imports of forty tons of germ growth stock suggest that the arsenal could have amounted to the high hundreds, or even into the thousands, of bombs. Had the Gulf War not intervened, Israel could have found itself by 1996 facing 500 SCUDs with biological warheads on them. This estimate of the Iraqi effort is not exaggerated. Baghdad bought 819 SCUD missiles from the Soviet Union to fire against Iran. In one six-week period in 1988 it fired 189 rockets at Iran's cities. Disappointingly, from Iraq's point of view, only about 2,500 people were killed. The lesson Iraq drew was that a deadlier warhead was needed; along with parallel programs in nuclear and chemical weapons, the development of biological weapons was pushed to meet this need for greater killing power.

The West treats biological warfare as a problem of nonproliferation. The Iraqi program is discussed in the language of arms control, of inspections and verifications. Much effort is applied to arcane problems such as "dual use," the confusion arising from the fact that the same equipment that makes germ weapons can also be used to brew beer. Sanctions are imposed; UN special commissions are established; restrictions are placed on the sale of biotechnology.

The significance of biological weapons is not that they are "diffi-cult" from the standpoint of arms control. Rather, their significance lies in the way they break the grip of established military powers. Strategic conditions are very different in a world with biological weapons than in a world without them. They open new paths of mil-itary evolution, just as the steamship and the tank once did. But they are much easier to build. Countries held back by lack of expertise in fields such as electronics or aeronautics can always find chemists and biologists. This possibility introduces major uncertainties for West-ern political leaders, military planners, and intelligence experts who have little experience in forecasting a future not dominated by their own well-understood conventional weapons.

For the moment, suspend this nonproliferation perspective and look at biological warfare instead as a business: a small underdevel-oped competitor tries to spread its influence at the expense of big-ger, richer, more advanced countries like the United States. The viewpoint is like that of a start-up software company (the early Microsoft) taking on an industry leader (IBM) with deeper techni-cal knowledge and a larger research program. Competing head to head against such an adversary is hopeless, so an alternative field of competition must be found. Chemical or atomic weapons are both good candidates, but we shall focus for now on biological weapons. Also, Iraq is the country that got caught, so more is known about its effort. But many Asian states (China, Iran, Syria, Pakistan, North Korea, others) are known to have biological pro-grams. In addition, Russian biological warfare experts are known to be selling their services throughout Asia.

Germ warfare is a disruptive technology because it upsets key Western advantages and negates its know-how in fighting a con-ventional war. It subverts the Western military structure by attack-ing its vulnerable rear area, the soft inner core of administration, logistics, and support, along with population centers. This nullifies the great Western advantages in smart weapons and electronic intelligence on the front lines. If the Gulf War is thought of as the prototypical modern war, then biological warfare is the prototypical postmodern war, one that negates the conventional advantages of the modern Western powers.

The political sensitivities of Moslem countries such as Saudi Arabia or Qatar require Western military bases to be cordoned off; hence, they are relatively large and few in number. This drives up the West's dependence on each base and explains why in the Gulf War there was such a mountain of supplies at each one. Everything had to be stocked in one or two locations. This requirement also conforms, however, to American logistical practice. It is easier to move supplies around one big base than among several smaller ones. With many smaller bases, an enormous intra-theater transportation system would be needed to ferry supplies and equipment back and forth, a system even the U.S. military does not possess. With larger bases, the only requirement is for long-distance haulage to move equipment and supplies from the United States. Once unloaded at the bases, equipment can be sent directly into battle. Note that this practice continues a centuries-long tradition of Western countries operating in Asia's harsh terrain by limiting the transport of ammunition and supplies to carefully selected compounds.

Dependence on a few big bases greatly increases vulnerability to biological attack. The SCUD is not an especially accurate missile, but it is accurate enough to hit a large base. Bases are area targets, and the SCUD is an area weapon. The great administrative and technical intricacies of the Western military effort are not seen at the tip of the military spear, but on its shaft. The tip—the frontline units—needs ammunition, fuel, and spare parts for its complex vehicles, radars, and radios. Biological weapons attack the shaft, the long, vulnerable part of the Western spear, comprising trucks, parts inventories, repair shops, and air and sea transportation. The shaft of the spear is not nearly as well protected as the tip is. In combat, airplanes and tanks are sealed off from the outside world to protect their operators and their inside electronics and mechanics. But on bases, motor pools and ammunition stocks are unprotected. Operators need freedom to move about without encumbrance in their resupply and maintenance work.

In Clausewitz's terms, biological weapons move the center of gravity of the battle from the front, where the United States has advantage, to the rear, where it does not. To "harden" the supporting "shaft" on the spear is technically possible, but extraordinarily expensive. It would entail an enormous reallocation of resources away from the spear's tip, a difficult trade-off for defense planners to contemplate. Germ weapons target American lives, not planes and tanks. While the subject is a complicated one, the mere loss of equipment is something that is unlikely ever to restrain the United States in war. But the prospect of significant loss of life is a different matter altogether, more so now perhaps than at any time in the past.

This center of gravity would have shifted even further had Saddam Hussein succeeded in his plans. Iraq was building the ability to target not only American soldiers but also Israeli civilians and, figuratively, American leaders and the public. And he would have done this in a way that violated the norms of Western warfare, which for historical and cultural reasons hold killing with biological weapons to be an especially gruesome and repugnant method of war. Western conventional weapons are meant to destroy in a blinding explosion of accurately directed chemical energy. The ideal is instant demolition. Biological weapons, in contrast, stretch out the death agony for days or weeks, or even years. Whether anything such as death in war can really be described as "clean," the West's view is that raining bombs on Iraq was a "clean" campaign. Biological weapons are by their nature "dirty." They are outside the bounds of accepted behavior, and in this they change the game of war to something much more horrible.

Innovation occurs in biological warfare, even though it tends not to be recognized in the West because the subject is abhorrent. This is a very shortsighted view. Mixing readily available anthrax and botulinum can result in a whole new compound of terrifying lethality. Such innovation is cheap. It doesn't demand an army of highly educated scientists. Unconstrained by American scruples, Iraq could gather data from tests on live prisoners, and it probably knows more about the effects of these weapons on people than the United States does. The United States actually knows very little about this subject because human testing has never been conceivable. But in Iraq such tests are known to have been conducted.

Germ weapons are indiscriminate not only about whom they kill but where they kill. They depend on the vagaries of wind and rain, which can steer the killing clouds in random directions. In the West this is seen as a major drawback. Integrating biological weapons into conventional forces goes against every precept of predictability and measurability that drives the Western approach to war. This is why germ weapons are so often dismissed as mere terror weapons by Western experts. But this very observation illustrates a larger difference. It compares germ weapons to the precise firepower of instant destruction preferred in the West. Measured this way, of course, they don't stand up. But from the challenger's point of view, not scoring well on such standard measures is not a bad thing.

A challenger, whether Iraq or Microsoft, knows that by competing in conventional ways it is bound to lose. It also understands that its adversaries are reluctant to change the rules. Dominant players would much prefer to keep the contest confined to areas where their well-developed skills work to maximum benefit. Confronted with an unconventional challenger, the leaders resort first to denial: "German tanks cannot possibly break the Maginot

Line." "Poorly equipped Viet Cong are no match for U.S. fire-power." "A software company cannot possibly knock IBM out of its domination of the PC business." Then, if the threat mounted by the newcomer proves effective, the dominant player starts to see that its advantages are not so great after all. At this point panic sets in. Morale in the French high command disintegrated. Washington saw a disaster in Southeast Asia that it could not escape from fast enough. And IBM shuffled its executives.

Biological weapons, like many disruptive technologies, are the very opposite of Western armaments. They are directed at area targets, not point targets. They operate in the rear area, not at the battle front. They go after the most vulnerable parts of the system, its logistics and support, not frontline weapons. Biological weapons are design to kill people, not things. They are "dirty" weapons, not clean ones. Accuracy and speed, the touchstones of American military innovation, are largely irrelevant in the field of biological warfare, where specialized command and control arrangements and strategies for use, such as brinkmanship, count for more.

COMPETING WITH DISRUPTIVE TECHNOLOGIES

The military modernization of Asia is taking the course of disruptive technologies—specifically, the development of ballistic missiles and weapons of mass destruction. Such technologies are transforming military geography and making U.S. bases in Asia, its "forward bases," vulnerable. The problem is not one of rogue states trying to make trouble. Nor is it that China is challenging the United States, unless possessing a modern military itself constitutes a challenge. Rather, a broad structural change in the technological environment of defense is taking place. Against peasant armies—even if upgraded by advanced tanks and artillery—the United States held the technological advantage. Against peasant labor, the United States had military capital. Washington could increase the killing power of its weapons to offset Asian numbers. But Asia no longer squanders human life in peasant charges against Western firepower, as Mao Zedong, Kim Il Sung, and Ho Chi Minh once did. Economic growth makes life more precious, just as it did in Europe. In place of enormous armies forging national identity through struggle against colonial and capitalist oppressors, advanced technology weapons have become the hallmarks of successful statehood and the symbols of progress.

Developing countries have always used Western military technology. But until the 1980s they imported it directly from the West without indigenous development. The shah of Iran and others were willing buyers of the West's latest military toys. There was little concern about these purchases because poor training and a lack of maintenance greatly limited their effectiveness. The shah had a thin overlay of modern Western equipment, a patina of a modern force, atop an unmotivated and undisciplined army. Most recognized this for what it was, a rump army propped up by Western technicians who could pull the plug at any time. The pattern held in most other countries that imported their armaments from the West. The abrupt collapse of the shah exposed this sham. Like the army of South Vietnam, when his army was called on to act, it melted away. All of the modern equipment imported from the West—the jets, the tanks, the guns—went for naught. The Western firepower of Iran was built on a foundation of sand.

The lesson was reinforced in the Gulf War. The Iraqi army fell over itself trying to surrender to the Western coalition. The conclusion many drew was that developing countries could not simply adopt the weapons of the West but needed a much deeper social and economic transformation to use these weapons in the way a Western army would. The entire society needed to modernize, not just its weapons. It was also said that totalitarian governments in particular did not have the "openness" needed to develop the modern fighting force, even though the Soviet Red Army, the Wehrmacht of Germany, and the Viet Cong were hardly "open" institutions, yet they performed rather well militarily. If only massive social transformation could produce an effective fighting force; then developing world military forces were many decades away from standing up to the West.

But while the shah's Iran fit the old model of importing from the West, Iraq, China, North Korea, India, Pakistan, and Syria were taking a different path. Rather than shopping for the latest in Western armaments, they redirected their efforts toward developing disruptive technologies. Iraq broadened its germ warfare program, North Korea started its missile buildup, China began its short- and medium-range ballistic missile program, and Indian nuclear scientists started to make small nuclear bombs for its missiles and air force.

The West continued to pursue its sustaining technologies, weapons that reinforced its military advantages. Cruise missiles were made more accurate, tanks faster, communications more intricate, and aircraft more stealthy. But in Asia the shift was from sustaining technologies imported from the West to disruptive technologies that overturned the West's main advantages. The contest was shaping up as one between the sustaining technologies of the West and the disruptive technologies of Asia.

One effect was to make fixed military bases an endangered species. Biological weapons make bases vulnerable; in combination with ballistic missiles, they present an overwhelming threat to bases. Missiles almost always get through to their targets because their speed makes them hard to shoot down, much harder than an airplane. When tipped with atomic warheads, or biological or chemical ones, the danger is made all the greater. Their disadvantage—lack of accuracy—no longer matters as much.

Disruptive technologies have changed the military game in Asia because they enable Asian countries to try to play catch-up with the West. They deny the West's assumptions about its capabilities, such as the freedom to fight from advanced bases, the ability to move forces to Asia without danger of harassment, and the ability to restrict a war to conventional forces confined to a small geographic box without the risk that it will spill over to a wider area.

ZONES OF EXCLUSION

Ballistic missiles break down the entire strategy of forward engagement from fixed bases. They are directed at the key vulnerability that Western powers in Asia have always faced, but that until recently Asian nations could not exploit. The effect of the current shift to missile forces in the belt of countries from Israel to North Korea is not that they will be launched against American bases at the first opportunity. The more interesting and important transformation is spatial: the shift is creating strategic exclusion zones and turning what historically were bases for projecting influence into Asia into places of high danger and political sensitivity.

Images from the Gulf War tell a revealing story here. Although a great deal of attention was given to the super-accurate bombs and missiles of the United States, it was the freedom to concentrate military power from around the world that was the real key to victory. U.S. jet fighters, tanks, and helicopters require high levels of maintenance and enormous logistical support. In the outside world—the civilian world of CNN and PBS—the enduring image of the war is one of U.S. laser-guided missiles landing right on the roofs of Iraqi buildings. It was these that led commentators to compare the Gulf War to a computer video game. The reputation for U.S. military invincibility was fed by these video clips.

On the inside, that is, among the Joint Chiefs of Staff, the image that symbolizes the Gulf War is much more prosaic. It is the photograph showing the mountain of supplies built up in Saudi Arabia between August 1990 and January 1991. An immense store of arms and munitions was amassed during the six months before the shooting started, a testament to America's ability to move equipment halfway around the world to the middle of a desert. It allowed the United States to be a military superpower in a place that initially lacked water or shelter. But it required a big base to house it all. Without bases, there can be no concentration of military power. Weapons cannot be stored, let alone massed for use. No bases, no weapons. It is America's singular military weakness in Asia. The different weapons that can be used against U.S. bases are well known, of course, but are rarely viewed as parts of a bigger whole: a pattern of disruptive technology aimed at the West's military freedom to operate at will in Asia.

Nuclear Weapons

Nuclear weapons destroy bases, ports, and airfields with a single blow. Getting them to the target can be the hardest part of their delivery, because placing them aboard aircraft makes them liable to be shot down. The ballistic missile reduces this risk considerably.

Asian nuclear weapons make military intervention a much riskier proposition for the West. The danger, however small, of a nuclear attack against a base or an allied city is likely to translate into caution about provocative military operations. Nuclear weapons create distance between the United States and its Asian allies because they raise the stakes for U.S. intervention. Atomic bombs, because they offset the vast superiority of U.S. conventional forces, are the premier disruptive technology at work in the world today.

Chemical and Biological Weapons

Chemical and biological weapons are the disruptive technologies of choice for countries that are unable to cross the nuclear threshold or are looking for a supplement to their small nuclear arsenals. If successfully delivered, these weapons can shut down a military base for a long time, owing to the difficulty of decontamination. The mere possibility that these weapons might be used can force extraordinary burdens on the United States—requiring troops to be inoculated, outfitted with respirators, dressed in rubber protective clothing. And if these measures were to fail, the prospect of Marines vomiting, gasping, and writhing in agony would give any American president plenty to think about before putting forces in harm's way.

Cruise Missiles

Cruise missiles are small, unmanned, aircraftlike vehicles that can fly below the radar of ships or the warning radars used to defend airfields. China has cruise missiles and has sold them to Iran. Many other countries have shown an interest in them. Highly accurate, they can strike targets without any risk to a pilot and can be armed with nuclear, chemical, biological, or conventional warheads. They are useful for attacking ships, since they skim low over the sea, and for this reason they pose a major threat to any navy operating in confined waters such as the Persian Gulf. During the cold war U.S. ships had less reason to approach coastal areas. On the open ocean, against an enemy that does not possess a blue-water navy, cruise missiles are less of a threat. Forward engagement brings ships closer to land, where they are vulnerable to cruise missiles fired from the coast or small patrol boats.

Sea Mines

Mines receive little public attention, but they have become more dangerous for the same reason that cruise missiles have. Ocean depths are too great for most sea mines. But near the coast or in relatively shallow waters like the Persian Gulf, mines can be laid on the sea bottom, where they are hard to find. Old-fashioned floating mines, like the ones seen in World War II movies—big metal spheres studded with spikes—are easy to locate and destroy. Bottom mines stay anchored on the sea floor until they fire rocket-propelled charges up at a passing ship. The ship does not even have to pass directly overhead if the mine has sensors that can direct it to its target. China and Russia make these mines and sell them to many countries. They are difficult to locate and, once found, difficult to remove, sometimes requiring that frogman teams physically take them apart. Sweeping these mines is often compared to cutting the grass on a football field with a manual push mower. It can be done, but it takes a long time.

Space Satellites and Drone Aircraft

Satellites are the eyes and ears of a modern exclusion zone. They can spot reinforcements moving into Asia by tracking military aircraft and ships as they approach their forward bases. Drones are miniature airplanes with cameras and other sensors aboard that fly for eighty hours or more. They are cheap and can monitor wide areas. It could eventually be possible to attack surface ships with missiles using these systems.

The combined impact of these technologies will be to change Asia's military geography as fundamentally as it changed when the airplane made the maritime colonial empire obsolete. Whether it takes five years or fifteen is irrelevant. Military access to Asia by outsiders will be far more difficult in the future than it has been in the past. The National Defense Panel, a Pentagon advisory body, recognized the danger in its December 1997 report "Transforming Defense: National Security in the Twenty-first Century." The panel's description of the future security environment states:

> Even if we retain the necessary bases and port infrastructure to support forward deployed forces they will be vulnerable to strikes that could reduce or neutralize their utility. Precision strikes, weapons of mass destruction, and cruise and ballistic missiles all present threats to our forward presence, particularly as stand-off ranges increase. So, too, do they threaten access to strategic geographic areas. Widely available national and commercial space-based systems providing imagery, communication, and position location will greatly multiply the vulnerability of fixed and, perhaps, mobile forces as well.

This trend affects the United States more than any other country, because of its status as the world's sole superpower. Formerly safe areas around the Asian perimeter are becoming dangerous to operate in. Space long "owned" by Western forces is now contested by Asian powers themselves. Asia's perimeter lands and seas are now overlaid by the tracking and reconnaissance grids of a half-dozen different nations. China can point its surveillance system to the waters around Taiwan, where in the past it was for all practical purposes blind. Iran can monitor the sea lanes in the Persian Gulf, while India keeps watch on the Indian Ocean.

THE LEARNING CURVE

This emerging military situation tends to go unrecognized because the early performance of disruptive technologies is usually very poor. Ballistic missiles in the hands of a developing country start out as unreliable and inaccurate. Many blow up on the launch pad. Throughout the 1980s and well into the 1990s the standard joke in American intelligence circles was that they posed a greater threat to their developers than to any foreign enemy.

After India's nuclear tests in the summer of 1998, likewise, evidence that some of the shots were duds was taken to mean that the Indians couldn't pull off a modern atomic bomb effort. The overwhelming tendency in the West is to see the weaknesses and failures in Asian military modernization. This has long been the Western view of Asia: that it is incapable of importing Western arms to good effect because it lacks the social and organizational skills to operate the equipment successfully. At one time Americans took a similar view of Asian businesses, that they would never become competitive threats to Western firms because they lacked professional management skills.

Disruptive technologies are especially prone to being underestimated by Westerners because they often present an unfamiliar package of performance attributes. Consider missiles. It is well known that Iraq, North Korea, and others have tremendous interest in missiles. Certainly, a great deal of attention is given to missiles by the U.S. intelligence community. But using a Western rational framework, it is difficult to spot advances that do not fit standard profiles.

Ballistic missiles are an ideal platform for innovation because changes are easy to make. The Iraqi missiles fired at Israel and U.S. bases in Saudi Arabia during the Gulf War were known in Iraq as the Al Husayn missile, an upgraded version of SCUDs bought from the Soviet Union. But the Iraqis improved the Soviet SCUD's 180-mile range to 400 miles. In the United States an entirely new missile program would have been started to replace a 180-mile missile with a 400-mile one. New technologies would have been added, with many tests. What Iraq did was simply cut up three SCUDs to get the parts and metal shell needed for the longer-range Al Husayn. The warhead weight was reduced so that it could be carried on the new missile.

With these design changes, the missile had very poor accuracy and very little punch. Western intelligence simply could not believe that the Iraqis, or anyone else, would make such a weapon. Only in early February 1988, after Iraq fired 180 Al Husayns at Iranian cities, did it become clear that Iraq had built a long-range version of the Soviet missile. Despite its inaccuracy and small payload, it had enormous strategic impact on ending the war with Iran, and it nearly brought Israel into the Gulf War three years later.

In April 1998 Pakistan tested its Ghauri medium-range missile; according to a congressional review panel, this test could not have been anticipated "on the basis of any known pattern of technical development." U.S. satellites had searched for the gantries that would hold up the rockets, for test flights, and for increased activity at known missile plants. But nothing was found. Pakistan had simply fielded the missile without the testing that would have been done in the United States. In July 1998 Iran launched a missile with a range of 800 miles, sufficient to attack Israel and U.S. bases in Saudi Arabia. This, too, was not anticipated. Earlier the CIA had declared such a launch to be unlikely for at least ten more years.

The development pattern of Asian missiles is hard to understand because they are looked at through Western eyes. Iraq, Iran, Pakistan, and North Korea simply went ahead without the elaborate safety, testing, and scientific controls that would be standard practice in the United States and that would tip off observers to the existence of a program. Iran and North Korea have built underground construction plants for their missiles, effectively hiding activity from overhead satellites.

Such procedures are in sharp contrast to cold war missile-building, with Washington and Moscow cooperating so that each could see the other side's forces. Western observers invariably extrapolate from their own history. There is growing concern, for example, over China getting an aircraft carrier, developing an oceangoing navy, and modernizing its air force to challenge the United States. But for China to play this catch-up game would be virtually impossible. It could easily take fifty years to acquire the needed aircraft, electronics, and crew skills. It would be so costly that even to start down this road would be a sign that Chinese leadership had taken leave of its senses.

But improving missile accuracy or building germ weapons is a different matter. This is a much cheaper way to shift the balance of power because it goes after key Western weaknesses.

GOING BALLISTIC

Chinese missile development is a case in point. In 1995 and 1996 Beijing used a series of missile tests to intimidate Taiwan. Research equipment was carried on the missiles, but their payloads could just as easily have been explosive warheads. The first tests were apparently precipitated by President Lee Teng-hui's visit to the United States in June 1995. China was angered by the implication of diplomatic recognition bestowed on Taiwan by President Lee's visit. The second tests were triggered by the Taiwanese elections of March 1996, during which President Lee seemed to flirt with the idea of Taiwanese independence. Beijing has long said that any declaration of independence could result in the use of force to return Taiwan to China.

The urgent question among U.S. observers was: what was the Chinese leadership up to with such missile diplomacy? What were they trying to signal? Did it represent the political leaders bowing to pressures from the army, and if so, what did this say about the future direction of Chinese foreign policy? Such questions are like the old practice of Kremlinology: fragments of speeches and other tidbits of information were pieced together to make long-term forecasts of Soviet behavior.

But what the tests really showed was the learning curve of Chinese missile technology. The first tests took place in July 1995, when the Chinese fired two DF-15 missiles daily for three straight days. The DF-15 (for Dong Feng, or East Wind) is a mobile missile with a range of 600 miles. It is launched from a trailer, which can be hidden from satellite or aerial reconnaissance in the same way that Saddam hid his SCUDs during the Gulf War.

The 1995 tests were not a success. Of six missiles fired, one had to be destroyed over China because of a guidance malfunction. Two others hit the far outer edge of a predesignated target zone. Intelligence reports suggest that for the three missiles that landed inside the target zone, accuracy was poor. The standard measure of missile accuracy is the *circular probable error,* the radius of a circle centered on the target within which half of the missiles will fall. Using this, the DF-15 in 1995 had an accuracy of 2.4 miles. With this accuracy, it would be ineffective with a high explosive warhead.

Only a nuclear weapon, or perhaps a chemical or biological weapon, would possess the necessary kill radius to compensate for the poor accuracy. When another test was conducted in March 1996, matters had changed considerably. Four missiles were launched at two target areas, rather than one, in the South and East China seas, one near the Taiwanese port of Kaohsiung and the other near the port of Keelung. This time all four missiles landed with near pinpoint accuracy.

How could accuracy have improved so dramatically in only eight months? First, it must be understood what an amazing achievement this was. In 1958 the Atlas, the first American long-range missile, had an accuracy of one mile. Since accuracy was such an important factor, a tremendous amount of research was devoted to improving it. Only by the early 1980s did the U.S. Minuteman III and Pershing II missiles get down to accuracies measured in hundreds of feet or less. Likewise, in the late 1950s the Soviet SS-6 missile had an accuracy of one mile. Only with their SS-18s and -19s in the early 1980s did the Soviets achieve accuracies of a few hundred feet. What took the United States and the Soviet Union twenty-five years to accomplish, China duplicated in eight months.

The DF-15 missile has what is called an *inertial navigation system.* This is an onboard computer programmed to keep the missile on a path predetermined before launch. Once launched, there is no further communication with the missile. With pure inertial navigation, there is no new information on whether it is on or off course. It is difficult, but not impossible, to achieve high accuracies unless the computer aboard is extremely advanced—comparable, for instance, to American or Soviet guidance technology of the early 1980s.

This requirement was thought to be beyond China's capacity in 1996. It is possible that the Chinese obtained such a system from Russia, or that engineers obtained the know-how there and built it themselves. Another possibility is that a midcourse correction was fed into the missile guidance system when it was in flight. An in-flight missile's position location can be obtained from navigation satellites. Then the programmed flight path can be compared with where the missile actually is.

Deviations between the two can be calculated by the onboard guidance computer, with the results fed to a small propulsion system to put it back on course. There are currently two space-based systems that allow this calculation, one American—the Global Positioning System, or GPS—and a Russian system called Glonass. China has entered into joint ventures with international firms to manufacture GPS and Glonass receivers. And it is known that it has a research program to link the DF-15 to navigation satellites. If so, this would be a much faster way to improve missile accuracy. Whatever the method, evidence pointed to Chinese confidence that accuracy had improved. The impact areas for the tests were moved from one hundred miles off Taiwan in 1995 to twenty miles off-shore in the 1996 exercise.

The implications of Chinese missile improvements are ominous for Taiwan. This island nation is extremely dependent on its critical infrastructures. Launching forty-five missiles with conventional warheads, China could virtually close the ports, airfields, water works, and power plants, and destroy the oil storage of a nation that needs continual replenishment from the outside world. This could be done with minimal civilian casualties, comparable in size to attacks the United States has launched on three occasions against Iraq since the Gulf War—attacks criticized as "pinpricks" because of their small numbers. The difference is that the United States was not trying to shoot down Iraq's economy.

SUPERPOWER LITE

It is important to appreciate just how precarious the U.S. military position in Asia is. The large U.S. military presence there is concentrated in a small number of bases, while the bulk of U.S. military forces are back in North America. Bases in Turkey, Saudi Arabia, Kuwait, Qatar, Diego Garcia, Guam, Japan, and South Korea are vital to span the immense Asian rim. Asia is four times larger than Europe, effectively limiting American military power to its periphery. Even so, over such a huge distance it is impossible for one set of bases to be used for all contingencies. Airfields in Dhahran, Saudi Arabia, would be useless in a Korean war, nor do prepositioned army stocks in Qatar help defend Taiwan. For this very reason, war plans during the cold war always called for much earlier nuclear weapon use in Asia than in Europe.

After the cold war Washington changed its geographic focus. It gave greater strategic importance to Asia, acknowledging peace in

Europe. Between 1988 and 1996 U.S. armed forces, active duty and reserve, were cut from 3.3 million to 2.4 million. But Asia was shielded from troop cuts; 210,000 troops were removed from Europe, 86 percent of the total reduction in active-duty personnel. Troop strength in the Middle East actually saw a small increase, while the Pacific underwent a token reduction of 32,000. Washington downsized from the cold war and reoriented its forces and strategy for Asia.

The political strategy accompanying these changes is detailed in a 1998 Pentagon document titled "The U.S. Security Strategy for the East Asia–Pacific Region." Cutting through the diplomatic language, with its support of democracy and open markets, the military rationale is essentially to keep doors closed that no one, at the moment, is trying to open. According to this report, the United States stays in Asia to forestall national competition among Asians; to reassure everyone that the United States will not allow the use of force to settle major issues like Taiwan or Korea; and finally, to be a substitute major power, which is preferable to an Asian country filling that role. The U.S. policy community has convinced itself that however much the Chinese may dislike America's presence, Beijing surely prefers it to the alternative of a rearmed Japan.

One critical part of the U.S. strategy, though nowhere mentioned in the Pentagon document, is essential to making it acceptable at home. At a time of scant public interest in foreign affairs and concern over the federal budget, forward engagement can be had on the cheap. Defense budgets have been trimmed back in the 1990s, with no Soviet Union left to fight, but no other country has emerged to challenge the United States as a superpower. Global leadership is an irresistible bargain when it takes up a smaller share of GNP each year.

But the U.S. forward engagement strategy was forged in the comparatively benevolent world of the early 1990s. A curve in the shape of a broad ∪ describes adversarial military power in the 1990s. The downward leg on the left side marks the collapse of the Soviet Union. The ascending curve on the right signifies the change from research programs to actual deployment of missiles throughout Asia. The price of global domination is about to go up, sharply.

The United States already has restricted freedom in its forward bases. Political strictures come from the need to obtain advance approval from host countries. In the Middle East, South Korea, and Japan, this can be tricky. Saudi Arabia and the Gulf states do not want to be seen as American vassals, nor can they risk the domestic political dangers of virtually conceding their sovereign territory to the United States. When the outside danger is great, as it was after the Iraqi invasion of Kuwait and has been several times since, they will assent to a buildup of American forces. But doing so requires frequent and costly logistical moves. Host countries, and Washington, have to make a commitment decision to move forces forward each and every time there is a crisis. This is altogether different from the cold war, when there was a tacit understanding between Washington and Moscow that both sides rushing forces forward would be a path to nuclear confrontation, so a mutual inhibition grew up about sudden movements of combat troops. This slower tempo of military action dampened escalation.

This mutual restraint has given way to a new dynamic. Forces are rushed to the front, in Iraq or Taiwan, whenever the political situation seems to call for it. The dynamic is now a sequence of crises, near-crises, presumed crises, and noncrises, each with a consequential political decision to move U.S. forces. These movements are carefully scripted by planners because they involve a staggering array of supplies and people coordinated to arrive in exactly the right order. But political consent is not so easy to plan. Host-nation politics may conflict with the tight synchronization required. A hitch in these movements—any kind of delay—is the nightmare of Pentagon planners. Then forces could be set to move, with nowhere to park at the other end!

This cycle of moving forces forward followed by hauling them back puts the United States in the position of constantly reacting to Iraqi provocation. One cycle influences the next, presenting opportunities for countries to try a little gamesmanship. Saddam provokes a minor crisis by tossing weapons inspectors out of the country or moving a brigade toward Kuwait. For the United States, each such occasion triggers a buildup of forces, costing a minimum of $2 billion, plus the political cost of going to the nervous Gulf states each time for permission to land. "Spoofing," or purposefully staging crises for the sole purpose of drawing a reaction, has become one of Iraq's favorite ploys. It is done to annoy Washington and to force U.S. leaders to cry wolf so many times that support among host nations and allies is undermined.

ARMS CONTROL AS STRATEGY

This game will become even trickier as more Asian nations acquire ballistic missiles. With missiles, the country provoking a U.S. deployment can threaten to hit the landing areas of incoming U.S. forces. U.S. strategy in Asia depends on arms control to forestall this from happening. It creates a fascinating juxtaposition. A legalistic Western concept, arms control by treaty, is the means by which the West attempts to maintain its monopoly of the technology used to dominate Asia. If negotiating with a Westernized Soviet Union on arms control was difficult, doing so with the many countries of Asia to preserve this advantage promises to be even more so.

The United States has pursued with great energy an agreement known as the Missile Technology Control Regime. This thirty-one-nation agreement bans international trade in critical components needed for missiles, navigational computers, and guidance systems. It is one of the principal tools used to retard the development of missiles in Iran, Iraq, Pakistan, North Korea, and Syria. Its success is ragged, however. Russia and China sell missile technology with near-impunity, and the Missile Technology Control Regime has developed serious leaks. Violators like North Korea flood the marketplace with their missiles. In addition, recent interpretations of the agreement allow export of space-launched rockets, of a kind used to put weather and communications satellites in orbit. Since there is virtually no difference between space-launched rockets and ICBMs, an absurd situation results. Obviously to disguise their true military purpose, countries are now building space exploration rockets with short-notice firing systems on mobile trailers.

There have been successes in arms control. Beijing has agreed to the Comprehensive Test Ban Treaty, which bans all testing of nuclear warheads. Yet thinking about arms control in terms of a scorecard of successes and failures misses a larger point.

Arms control is now being asked to sustain permanently an asymmetric advantage. By locking countries into a low state of military development, the United States can maintain its superiority on the cheap. There is a tradition of using arms control in this way, a tradition neglected during the cold war when arms control was directed toward stabilizing the balance of terror or, more rarely, reducing it. In the 1920s the United States, Britain, and Japan signed an arms control agreement to limit battleship construction. Its purpose was to limit the Japanese navy so that Britain and the United States could keep their global military leads without paying for new battleships. It was the first time that Western nations colluded to constrain the military modernization of an Asian industrial power.

Using arms control for one-sided strategic advantage still allows a country to argue that such agreements are in the general interest of mankind and international peace. Beating swords into plowshares is a much more appealing argument than cold strategic logic. Placing arms control at the center of diplomacy also makes it seem that a new cooperative security system is emerging in which countries will refrain from threatening one another. Yet the reality is not quite this simple.

Preservation of the asymmetric situation whereby the greatest military power in Asia is not Asian depends on arms control. But China will never sign an arms control agreement that makes U.S. bases in Asia invulnerable. Nor would Iran, Iraq, North Korea, or even Israel give up their missiles. The one-sided character of such an agreement, the appearance that it permanently locks China and others into technological inferiority to the West, would be clear to all. Failure to reach arms control agreement thus undermines America's whole Asian strategy.

The stage is set for arms control to take on a greater significance even than it did in the cold war, when the objective was stabilization rather than one-sided advantage. In recent years there have been many diplomatic disputes over sales of sensitive technology by China and Russia to nations in the Middle East and over Chinese sales to Pakistan. These disputes are a pale harbinger of what is to come, for their importance will increase as the costs and problems of defending forward bases become impossible to ignore.

AMERICA'S MAGINOT LINE

If forward engagement is to mean anything, it requires credible military power on the Asian periphery. Yet the foundation of that power, forward bases, is eroding. The result—only somewhat masked by the triumphal rhetoric of cooperative engagement and strategic partnership—is a deterioration of the long-term capability of the United States to do what it says it is going to do. The United States is reacting to these developments with a major initiative in defensive missiles, that is, missiles that can shoot down ballistic missiles. Missile defense has been around a long time. But in the past it has focused on national missile defense, protection of the U.S. homeland from Soviet attack. Now, in an attempt to defend forward bases, anti-missile missiles are needed to protect them.

Protection of forward bases is likely to produce an American Maginot Line, a defensive effort to stay in Asia for another fifty years by protecting the bases of a bygone era. Like the original Maginot Line, it could easily produce a defense psychology as well, a posture of increasing caution and restraint. Forward bases could become like the Crusaders' forts of the Middle East—heavily defended, isolated outposts with little effect beyond their immediate perimeter.

Missile defense is a lot like hitting a bullet with another bullet. The difference is that missiles travel faster than bullets, and they explode. To be effectively killed, an incoming missile has to be destroyed far from its target. Warheads travel so fast that they can still do great damage when armed with nuclear or biological warheads after being hit because momentum will carry even a damaged missile toward its target. Hence the daunting problems of defending against ballistic missiles: the speed with which they travel, the short time in which to get a kill, and the need to demolish the warhead far from the target.

It is difficult for nonspecialists to imagine the kinematics of a missile–anti-missile engagement. Stopping a missile is altogether different from interdicting a car. Saddam's Al Husayn and the Chinese DF-15 fly between 5,000 and 7,000 feet per second at their maximum. An object traveling at sixty miles an hour is moving at 88 feet per second. A shotgun fires at about 900 feet per second. The U.S. army's M-16 rifle, designed for high-velocity fire, fires rounds traveling 3,200 feet per second. A ballistic missile travels twice as fast as a bullet from an M-16.

It is most efficient to kill these things near their launch point. Then, only a few defenders are needed to cover a large area. An airplane with an anti-missile or high-energy laser aboard—a space-based missile shooting down at the earth—or a naval ship close to the firing point provide efficient geographic coverage. Otherwise, a large number of point defenses protecting widely spread targets are needed, and the cost of the defense goes up. This geometry means that the target, sensors, and anti-missile killers are widely dispersed, spread apart over the earth's horizon. A complex system of space satellites and computers is needed to track the attacking missile and bring the whole defense together. At the altitudes where missiles fly, the air is so thin that blast waves are much attenuated, so an interceptor can't simply blow up in the vicinity of its target missile; it has to hit it directly. Weight considerations argue for a fast, lightweight missile, homing on its target by infrared (heat) sensors, smashing into it at such high speed that it causes disintegration. This is a technological problem still awaiting solution.

National missile defense of a kind considered throughout the cold war, most prominently in President Ronald Reagan's Strategic Defense Initiative (SDI), is an easier problem than theater ballistic missile defense. ICBMs launched from Russia have a flight time of thirty minutes to most targets in the United States, compared to five to eleven minutes for a typical short-range missile. The longer time allows interceptor missiles to reach sufficient altitude to take on the incoming warheads before they start to close on their targets. With eleven minutes or less, that is barely possible, and then only if everything happens almost instantaneously. The defense must have a tightly scripted response of virtually automated actions. The smallest delay or malfunction can undermine the mission.

In 1993 an important test of defensive missiles showed the high level of anxiety in the Pentagon about ballistic missile attack on forward bases. Chemical weapons, like the notorious VX nerve agent, are best delivered not as a single bomb package, but as a bundle of small bomblets. Syria's SCUD C missile, for example, uses minicanisters filled with nerve gas packed into a warhead. Warheads can be built to cast these canisters over a wide area. For contaminating a base, port, airfield, or any area target, this is exactly what is needed. In the 1993 test a missile carrying thirty-eight canisters of water was used to simulate a ballistic missile armed with such a chemical warhead. It was annihilated by the defender missile.

But that didn't stop the debate over whether theater ballistic missile defense will work. Some critics charge that tests are conducted to minimize difficulty, that in effect they are rigged for success. Others counter that experiments like the 1993 test demonstrate that it is not impossible that these technologies could shoot down missiles. Everyone agrees, however, that the problem is extremely challenging: the smallest glitch could mean failure, given the high speeds and short reaction times.

Whether ballistic missile defenses will work depends on technical specifics and geometry. American technological know-how could fare well here—our defenses might even be good. But there are many issues that have not even begun to be thought through, complications and details that illustrate a larger point: missile defense is in part a technical problem, but it is also a political and economic one with a specific geographic shape dictated by the fact that American forward bases ring the Asian coastal rim. A principal rationale of this system is cooperative engagement, but the missile defense issue shows a major disconnect. If our bases are so welcomed throughout Asia, why are so many countries building forces that will make them obsolete? Missile defense has not come to grips with this problem, and it is fundamental.

For thirty years every administration that has faced the national missile defense question has done so recognizing that the only way for it to work is to limit the size and character of the threat. Whether an administration favored missile defense or opposed it, all saw the imperative of getting a handle on the offensive side of the ledger. If offensive threats were not limited, defense would have to invest vast resources to counter them. This is the core economics of the missile–anti-missile game.

In the late 1960s Secretary of Defense Robert McNamara initially opposed missile defense for the United States because the Soviets could easily counter it by building more missiles, essentially exhausting the defense with cheaper offensive missiles. The Nixon administration initially favored defense, but only as arms control negotiations were simultaneously pushed to limit the enemy's offensive capabilities. This produced the Strategic Arms Limitation Talks, known as SALT. Likewise, the Reagan administration advanced its Strategic Defense Initiative along with the Strategic Arms Reduction Talks (START), again an effort to get a manageable handle on the offensive threat. These treaties worked for the purposes intended; over a period of some thirty years Moscow and Washington put limits on their offensive missiles. The two cooperatively engaged.

But there is no strategic partnership between the United States and Asia on missiles. Neither China nor India is involved in the START talks, or in any talks that might follow it. India and China have gone out of their way to argue that until the United States and Russia cut their nuclear arsenals much more deeply than they have, there is not even any point in bringing them into negotiations. With the recent Clinton administration decision to rewrite the ABM Treaty, the chances are even thinner that any agreement on long-range missiles will be reached.

Nor is there any plan to limit shorter-range missiles. The potential contributions of a revolutionary new arms control initiative are easy to identify here, but difficult to accomplish. Washington could push for limits on medium- or short-range missiles. Just as there is a call for a ban on land mines, so could there also be, in theory, a ban on at least certain missiles. The current Missile Technology Control Regime doesn't try to do this, for it limits cross-border trade only in missile parts, not in missile possession. Such a revolutionary new arms control plan would be exceedingly difficult to verify, even approximately.

Without some limit on the number of offensive missiles, defense becomes exceedingly expensive, partly because of the nature of disruptive technologies, which are platforms for innovation. Thus, China compressed twenty-five years of missile development into eight months. Missiles can be made to fly faster by streamlining their warheads, thus forcing a defender to retrofit its own force to counter it. They can be made to spiral in a path that is difficult for an interceptor to hit, or they can carry decoys to distract the defender's missiles. The number of options is enormous and nearly always cheaper than the cost to the defender of coming up with a response. If a faster warhead is built with less chance of detection by radar, for example, it necessitates a new suite of radars and space tracking systems to find it.

This is the significance of Chinese espionage against American nuclear weapon labs. It is not the spying that is significant, because all countries do that. But the object of the spying is highly revealing. Chinese spies sought the design for the deadly small warheads of America's nuclear arsenal. This espionage target was not some general collection of American technical data. Rather, it fit the clear pattern of Chinese defense modernization. With this design, China can put bombs on mobile missiles, submarines, and the ICBMs that can strike the United States. In short, China wanted a modern nuclear force of lightweight warheads. With these, whole new areas of strategic innovation are opened up.

With the stolen bomb designs China has now compressed two decades of nuclear weapons research into a few short years. Together, the ballistic missile and warhead improvements give China enormous flexibility in responding to American actions, on Taiwan or on any other issue for that matter. Is Washington sending anti-ballistic missiles to Japan and South Korea of a kind that could be transferred to Taiwan? No problem. All Beijing has to do is keep the plants running to field another hundred missiles, something that is now relatively cheap to do because the hard part, the knowledge of how to make them, is solved courtesy of the loose security controls of Washington's engagement policy toward China. Once the factories are up and running it is cheap to keep a second shift working, especially given Chinese labor costs. In 1995, China had about fifty missiles aimed at Taiwan. Now, 200 missiles are there. In a few years, a thousand missiles are likely to be pointed at the island. The mix of these missiles—the fraction that are short range and capable of hitting Taiwan, and those of longer range that can take out American bases in Japan and South Korea—is entirely up to the Chinese. By tilting the mixture toward longer-range missiles, Beijing not only turns America's Asian bases into hostages; she also inhibits U.S. power in the whole Pacific Basin.

Without size limits on the offensive force through arms control, the economics of defending forward bases becomes prohibitive, spelling the end of superpower lite. It can be done, but not cheaply, and certainly not with the defense budgets of the 1990s. There is an additional political anxiety. Political tensions are exacerbated when one party or another feels constantly behind in the game. Sooner or later, Washington is going to get irked at Chinese missile deployments that force a fresh round of theater missile defense expenditures. Nor should one discount the possibility of bluffing. During the cold war the Soviets leaked false information about some of their missiles, exaggerating their capabilities essentially for the purpose of infuriating and frustrating the Pentagon. It isn't hard to imagine Iraq or China doing the same thing.

Washington and Moscow considered missile defenses in a very

different context from what the United States might face in Asia. In the cold war we were protecting ICBMs in super-hardened, underground concrete silos. If the defenses allowed the ICBMs to survive for a few hours, they had done their job. Bases are called "soft" targets for a reason. Exposed troops, airplanes, equipment, repair stations, and bomb stocks would need a virtually impermeable defensive shield, far more reliable than anything offered by President Ronald Reagan's SDI program. Unlike national missile defense, the protective bubble would be called on against repeated salvos. The war of the cities in the Iran-Iraq conflict saw Baghdad stretch out its attacks over six weeks.

That the United States is even considering defenses for its bases is important and instructive. Forward bases could become like Crusader forts in the Middle East. These forts did survive for a long time. The United States could go down this road if it opts to look at the problem in the narrow terms of deciding simply whether defenses will or won't work. It might do so, despite all the technological problems. But the important issue is not whether some missiles can be shot down, but whether the United States can remain an Asian military power in the future in the way it has been in the past by building a protective bubble around its forward bases. It almost certainly cannot. To stay a relevant player in the Asian military game, more far-reaching and expensive changes will be required. Say good-bye to superpower lite.

It is worth emphasizing that the initiative in this game originated in Asia, with its missile programs. There is a new security environment, one of the United States *reacting* to Asian initiatives, not shaping them to head them off in the first place. An anti-missile system does not increase U.S. military power in Asia, it only allows it to remain—if the system works as planned. The reactive character of this strategy has not received the attention it deserves.

3   RESHAPING THE ASIAN MILITARY

Civilian leaders, not military generals, are behind the move to weapons of mass destruction. These weapons offer a way to redirect the bloated and politicized military establishments that have grown up in Asia. The generals oppose them as a diversion of money away from the traditional army. But rulers can use them to reshape their armed forces for the post-postcolonial era, when giant land armies are more of a burden than a benefit. Seen this way, weapons of mass destruction do not represent the lurch toward militarization that they are often made out to be. Instead, they are civilian attempts to extend the modernization of the economy to the military sector.

Iraq, Iran, and North Korea all have civilian governments. They are not democratic or elected, but they are civilian. Generals don't run these countries; they answer to the civilian rulers. This is also the case in China and most other Asian countries. Rulers use the military to stay in power. These regimes are nothing like the tin-pot military dictatorships often found in Africa and Latin America, where generals hijack control to install a junta to run the show. Even Iraq's ruler, Saddam Hussein, has used the power of the nation-state, of civilian administrators and engineering technocrats, to modernize his country's economy and military. He may share certain temperamental features with Idi Amin, the former military dictator in Uganda, but he is far more dangerous to the outside world than Amin ever was because he mobilized Iraq's national energies in ways that include, but go well beyond, pointing the gun at his own citizens.

Economic modernization requires civilian leaders to control the army organization. Big infantry forces were the key institutions in the postcolonial era after World War II. They were essential for keeping order and preventing outside powers from causing trouble during the early days of nationhood. This was a time of explosive social transformation, when villagers were moving to the city and peasants were taking factory work, when the greater danger was not war between states but civil war within the state.[1] The large armies of China, India, and Pakistan held those tensions in check.

But in most of Asia industrialization and urbanization are well under way, and the armed forces are no longer needed to see the state through this transition. Very little benefit comes from maintaining an oversized, poorly trained army. Public order is better kept by specialized paramilitary units and police than by the army. The generals only meddle in politics and business. The old army is also useless in meeting the external threats posed from other countries with their own weapons of mass destruction.

Militarism—the triumph of the generals over civilians promoting economic growth—is in retreat across Asia. Yet this is the very time when the spread of missiles and weapons of mass destruction has

---

[1] The *state* refers to the authoritative decision-making institutions in society and is unique because it is the ultimate regulator of the legitimate use of violence; the *nation* refers to a shared identity and feelings of attachment beyond allegiance based on region, class, ethnicity, or religion. The Kurds and Palestinians are nations, but not states. Many Indian scholars argue that India is a state but not a nation.

accelerated. Western observers are baffled by this. With the cold war over, and with economic growth the priority across most of Asia, what do these countries want with missiles and atom bombs? One answer is that weapons of mass destruction turn the military's focus outward, giving it a new role in national defense, and not coincidentally distracting it from meddling in politics and business. Paradoxically, if the United States could somehow increase the political influence of the military in Asia, it would slow the proliferation of nuclear weapons, for what the old generals really want is a better-equipped version of the force they had in 1970. But from the civilian point of view, that's an expensive way to prepare to lose the next war.

REMODELING THE ASIAN MILITARY

Armies reflect their strategic environment. In postcolonial Asia the army was an institution with which to forge a new national identity transcending the regional and religious differences that colonial rulers manipulated to stay in power. The army was an instrument of mass indoctrination, a giant school with a core curriculum of nationhood. It was not a professional fighting force, for the army did not spend its time perfecting its battlefield skills. Instead, it focused on projects that increased national cohesiveness, such as helping farmers bring in the crops. In the 1990s Asia had seven of the ten largest armies in the world. The reason for this was political, not military. Millions were trained to identify with the nation rather than with their ethnic group, tribe, caste, religion, region, or village. Although such local self-identification endures, it is transcended by something else: identification with the nation. Army size was dictated as much by this nationalization effort as by foreign threats. In China the army had a much more critical role to play in building revolutionary fervor than in countering foreign aggression. Chinese troops were sent to the countryside to build railroads and thereby inspire others with their Communist zeal. In India the army's role was less direct. It stayed out of politics but still served as a giant training camp for creating "Indians." That is the reason it got so large.

The Asian army, with origins (typically) in revolt against the colonial government, became politicized through its nation-building activities. Then there was little to stop it from entering the new business enterprises that followed decolonization. In the United States or Germany the armed forces do not own businesses or play in politics, at least not to anything like the degree they do in much of Asia. In China and other Asian countries, the armed forces were the king makers, and political leaders had to buy them off by giving officers control of business monopolies and sweet retirement jobs. In Indonesia the army ran the state oil monopoly. Officers were often paid with stock in the company, giving them direct ownership rights. In China the People's Liberation Army (PLA) got into tourism, toy manufacturing, textiles, and missile components. In South Korea retired officers were given key jobs in the big industrial groups. Under the shah, Iran's oil business was populated with military favorites.

Military politics became the rule rather than the exception in postcolonial Asia. In Indonesia, China, South Korea, Pakistan, and throughout the Middle East, gaining political office required getting the support of the army. It was the single most powerful institution in the country. Leaders had to manipulate it, block it, and co-opt it to keep their hold on power.

When this politicized army went into battle, it performed terribly. Huge casualties, uncoordinated supply for the troops, and gross misconduct were the order of the day. China lost one million men in the Korean War. The wars between India and Pakistan were comedies of errors. In the Arab-Israeli wars Israel was able to exploit Arab military incompetence to win stunning victories, leading to Moshe Dayan's famous answer to the question of why Israel was so successful in war. The secret, said Dayan, was to "fight Arabs." The Iran-Iraq War could best be described as two uncontrolled mobs attacking each other, and the Iraqi army performed no better in the Gulf War.

The only Asian army to hold its own against the West after World War II was the Vietnamese, against the United States. The Vietnamese were very tough and disciplined. Having stood up to China for centuries, they found the French and the Americans easy to deal with. The United States understood so little of this history that it assumed, quite incorrectly, that the Vietnamese would buckle to U.S. pressure.

China is a good example of the giant Asian army and its limitations. Under Mao Zedong, it was a tool for political infighting. In the Great Leap Forward, Mao exploited the chaos that his own plan had produced by using the PLA to restore order on terms beneficial to himself. Then followed the Cultural Revolution of the 1960s; with the PLA's power on the rise, Mao set the radical Red Guards against the military to foreclose the possibility that they would act against him.

Under Mao, the Chinese army was more of an internal political force than a professional body concerned with fighting the United States or the Soviet Union. Its disposition showed this. It was organized into thousands of small cells spread around the countryside and grouped into giant field armies. These armies almost never exercised as a whole because coordination was too hard, and because communication between units was so severely limited. The general staff, made up of Communist Party leaders, could order the army to take in the harvest and to build railroads. These things it could do. But it lacked the infrastructure and training to fight a war against the Soviet Union or to serve as a flexible instrument of foreign policy. In Mao's China, highways were never built, telephone circuits were never laid, and railroad lines ran not to the borders for offensive attack, but to the interior for protection from invasion. As a result, Chinese military strategy was extremely crude, largely consisting of the mass wave attacks used against America in the Korean War.

The Chinese army was a military dinosaur. It could not coordinate large bodies of men. Its geographical reach was limited by how far the infantry could walk over harsh terrain. This is not to suggest that simple strategies could not be effective. In the Korean War this army walked to North Korea, where it dealt the United States a stunning blow. The Vietnamese army was equally primitive, but it, too, beat the United States. However, the degree of innovation in China, Vietnam, and most other Asian countries was always limited because there was so little technology and infrastructure to build on. Most often, innovation consisted of reorganizing the army into subunits of different sizes or specialized functions. For example, China kept large field formations, while Vietnam split its force into a regular army and a guerrilla force, the Viet Cong. In North Korea the army was split between a regular infantry and commando units trained for sabotage. Such innovations were useful, but they all relied on using mass in different ways to beat the enemy.

Asian nation-states now are more established. There is no danger of a return to colonialism, and the danger of civil war has declined since the early days of statehood, when it was anyone's guess who would rule the country. As a result, the political process has matured in South Korea, China, and India. In other countries such as Indonesia, North Korea, Pakistan, and Iraq, political maturity is still in the future. Nonetheless, for all of these countries the need for a mass army has diminished. The politicized army has gone too far in its business ventures, creating serious problems of economic instability. In China and Indonesia military enterprises strangle the national economy. In Pakistan the army's internal blood feuds have carried over into foreign adventures in Kashmir and Afghanistan that have brought that country to the brink of chaos.

The problem throughout Asia now is how to get the army out of politics and the economy. A new kind of military is needed: one that is outward- rather than inward-looking; one that is professionally rather than politically adept; and one that is flexible rather than rigid. Asian armies need to be downsized and replaced with modern forces. Reaping little benefit from its massive force, China has been cutting the military back from a peak size of 5 million in 1980 to some 2.5 million today.

Other countries would like to change to a more modern force, leaving behind the massive peasant infantry divisions of the postcolonial era. But politics and size make this hard to do. Like any large organization, the armed forces are exceedingly slow to reform, especially when they are run by influential generals who are resistant to any reduction of their authority. The military clings to its role, and it is very difficult for civilian leaders even to penetrate military organizations, much less turn them into flexible instruments of foreign policy.

The record on reform is long on need and short on results. Increasing money for new equipment as a way to engineer change is the dream of every Asian general. But the generals' dreams do not include the new weapons of mass destruction and missiles. Rather, they would buy thousands of artillery pieces, trucks, and tanks. They dream, in essence, of reproducing themselves as a better-equipped version of what they were twenty-five years ago. Yet showering money on the army in this way would have little impact on their modernization. Inefficiency, corruption, and bureaucratic intransigence would so distort such an expensive program that in the end it would probably only line the pockets of the generals who oversaw it.

Spending more money this way also makes no strategic sense. China and India would hardly be more secure with one million men added to their armies, even if they had more howitzers with which to arm them. Such a force would be not only irrelevant to the foreign policy universe these countries now live in but financially ruinous to the government. However, spending money on programs that actually extend the international influence and reach of political leaders is a different matter. The problem is how to do it.

Remodeling the Asian military around weapons of mass destruction, missiles, and other disruptive technologies offers leaders a way to get their hooks into these organizations. Disruptive technologies provide a new framework on which to build a military for the twenty-first century. Civilian leaders can bypass the bloated military bureaucracy by using a new avenue for civilian control over the military: working through special units created to acquire these systems outside of traditional military command lines. The new organizations—the atomic energy commission, biological warfare research units, missile construction groups—are populated by technocrats and bureaucratic entrepreneurs. As a result, the influence of the traditional army, the key institution in the postcolonial era, is being displaced by technocratic enterprises that are more responsive to civilian rulers' ambitions—both good and evil.

These new groups enable strategy innovations beyond the crude infantry tactics of the past. They free rulers from the strategic limitations of giant land-based armies. The new capabilities do not just change the old military balance, although they certainly do this. They do much more, becoming game changers for those who understand the new strategies they open up. Saddam Hussein, and others like him, now have the opportunity to give tangible expression to their evil delusions. The East, despite a tradition of profound strategic and philosophical thinking going back to the writings of Sun-tzu, has been manifestly incapable of asserting itself militarily. Missiles and weapons of mass destruction end this condition. They provide a technology platform for new strategies that tap a much richer philosophical tradition than that of the narrow-minded infantry generals of the postcolonial era.

EYES IN THE SKIES, AND ELSEWHERE

The information revolution spreading around the world brings much more diverse sources of intelligence to the Asian military decision-making system. Satellites, fiber-optic communication lines, computer networks, and cellular telephones disgorge information that will transform civil-military relations in Asia. The new information technologies allow a quantum jump in performance for key parts of the armed forces. For other parts, they make little difference. Missile forces especially can be made highly effective and insulated from the inefficiencies that pervade other parts of the military. This goes unnoticed in most Western accounts, which usually focus on particularly glaring examples of backwardness and take them as emblematic of overall performance. In some areas, like jet aircraft or mechanized ground warfare, the Asian military is extremely backward compared to America or Europe. However, this assessment overlooks the role of new information technologies in making missile strikes and other tactics highly effective.

The anatomy of the Asian military, viewed as an information-processing unit, is quite different than when looked at as a mass army consumed with politics. Most observers still look at the Asian military as a bloated, unprofessional infantry force mainly interested in garnering larger budgets and preserving its privileged place in society, however crippled by internal rivalries and political divisions. But looking at Asian military forces as information-processing systems, the question is whether different parts of the military can smoothly work together. To the extent that information technologies can make this occur, new strategies become possible. The issue for each Asian nation is no longer whether General X has allies on the general staff, but whether there is a warning system that can collect information about the state of enemy missiles in enough time to fire its own. Areas of military competence are emerging, still partially obscured by the poorly organized infantry armies. If we think about civil-military relations in China too much as a struggle for power, we think too little about whether China can collect, process, and store information that allows entirely new kinds of strategies. While both perspectives are legitimate, information technologies raise the possibility of a degree of political control over missiles and weapons of mass destruction that was never possible with infantry armies. As a result, a view of the Chinese military as political power brokers is likely to be far less important in the future than it has been in the past.

China's military communication system demonstrates how technology changes the politics of control. Until very recently, China's leaders communicated to the armed forces by issuing orders that were forwarded unchanged down the hierarchy. Messages were not altered at any level, for example, by adding intelligence information. The message from the top was reinforced with endless repetition so that no one would fail to grasp its meaning. Under Mao Zedong, this practice reached absurd levels. Most people recall the scenes of millions of soldiers chanting the slogans of Chairman Mao while waving his Red Book at the cameras. This image captures the essence of the Chinese military communications of its day: crude and political. The army was subjected to mind-numbing repetition of the chairman's quotes from his Red Book. While this had the effect of drilling home Mao's wisdom and forging a homogenous worldview within the army, it had some major disadvantages. Only very general guidance could be given about what to do. The communication technology, which consisted of wall posters, hand-written messages, and loudspeakers, did not allow for anything more.

This flow of communication was one way, from top to bottom. Subordinate units had no ability to ask what to do when conditions did not match the general orders coming from the top. An army run this way is bound to get into trouble because broad general orders usually do not fit the messy conditions of the field. The result is a highly inflexible military organization. When the Chinese military consisted of 5 million peasants, when it had virtually no navy or rocket forces, and when it was focusing on internal order, it was possible to get by with such a communication system. But against Vietnam in 1979 it performed very badly. Field units were in the dark about where the enemy was, or whether reinforcements were coming, because no communication grid existed to collect and distribute this information. Battle units did not know what headquarters expected of them, which direction to advance, or whether to advance at all.

The communication system in the Chinese military today is quite different. If the old image was one of millions of soldiers chanting the Red Book slogans of the chairman, the new image derives from the Chinese desire to obtain advanced U.S. communications equipment. Now a U.S. "consultant" burrows his way into the American political establishment to loosen up the government's ban on exporting high-tech communications gear to China. Acquiring Western-built communications satellites, computers, and digital cell phone systems has now become the top priority of the Chinese military. They are bent on getting this technology because it allows them to control a more far-flung enterprise that is much more interested in tracking American ships and foreign military units than the old Chinese army ever was. The new gear gives the Chinese general staff an ability to give tailored messages that are specialized for different subunits and regional commands. They allow field units to communicate two ways, to kick decisions back upstairs when orders do not fit local conditions. The new communication system turns the Chinese army from a 5-million-man mob that was long on enthusiasm but short on tactical skills into the beginnings of a coordinated professional military.

The Chinese have built a web of communications intelligence collection stations along their borders to monitor their external environment. This, too, is a major change from the past, when the main tactic was to retreat into the countryside, withdrawing from the outside world. Information is now pulled in from the bottom from sensors and monitors in space and around the Chinese borders, rather than pushed down from the top on the basis of ideological orthodoxy.

THE RISE OF THE ASIAN MILITARY-INDUSTRIAL COMPLEX

A new military-industrial complex is rising across Asia, built around missiles, weapons of mass destruction, and other technologies. It promises to be a significant bureaucratic and political force in the future. The army used to be the key actor in Asia, but a more technocratic arm offers to far surpass it in influence, mobilizing not the millions that armies did, but the scientific and engineering manpower of the nation for the use of the rulers.

India offers a good example of the rise of the Asian military-industrial complex. In India there is a parallel development of atomic and hydrogen bombs, ballistic missiles, a space program, a navy with greater reach, and extended reconnaissance from aircraft and satellites. Each of these programs has an institutional momentum and a cadre of engineers and hustling bureaucrats making demands on politicians and the national budget. The new technocratic forms are overlaid on a gargantuan, inefficient, and increasingly underfunded army. The composition of these two structures, one old and one new, is not simply a case of the modern replacing the old but rather typifies what is seen throughout Asia: modernized pieces of the armed forces linked to older, inefficient parts.

India's first atomic bomb, tested in 1974, was no exception to this pattern. It was so big that it could be delivered only by a giant cargo airplane. For the next decade India's nuclear arsenal, if it could be called that, was made up of these oversized contraptions. This bomb's utility was entirely symbolic, and as far as is known, there were no plans for integrating it into military operations. It was a stand-alone product of Indian nuclear scientists without the slightest consideration of its military uses.

Rajiv Gandhi fundamentally changed all of this when he gave the go-ahead in the late 1980s for building smaller atomic bombs that could be carried on ballistic missiles. This revitalized the Indian Atomic Energy Commission, increasing its status and prestige. It also marked a new kind of mobilization—not of divisions, but of scientific and engineering talent. But this wasn't all. The bombs required shell casings and shock-resistant packaging to ensure that they would actually work when fired on the front end of a missile. The Indian Atomic Energy Commission knew nothing about weaponizing an atomic bomb, as this is called. It therefore had to reach out to the sprawling Defense Research and Development Organization. These two giant establishments had never worked together before. But in the early 1990s they cooperated on the bombs that were tested in the spring of 1998. The official communiqué from the Indian government announcing the tests, interestingly, was a joint statement from both of these organizations.

The result of this cooperation was a rising interdependence between the two agencies, and their growing influence on the future shape of science and technology in India. Along with India's space agency, which has launched commercial and military satellites, they have become powerful forces in Indian society. The Indian space program is building booster rockets for launching spy satellites. There are many commercial uses for these launchers, and there is no clear divide between the military and nonmilitary parts of the Indian space effort. India's space vehicles have the capacity for an ICBM, a missile that could strike the United States and Europe. India is also experimenting with a sea-launched ballistic missile (SLBM), and it is reasonable to imagine that the Indian navy will be deeply involved in this project, because it brings the navy into the center of a key national program and allows it to outflank the army, its longtime bureaucratic rival.

Members of the new technocratic bureaus are becoming an influential class in India, in the way that the leaders of weapons laboratories rose to prominence in the United States and the Soviet Union during the cold war. Also like the military-industrial complex in the United States in the 1950s, and like the Soviet weapon design bureaus, these Indian organizations now have a powerful interest in self-perpetuation. Hence, relations between the civilian and military sectors have moved from power politics to a more complicated interdependence. As defense becomes more technological and intelligence grows in importance and quantity, governments have to consult their techno-military staffs more often. To ignore their advice is to run the risk of falling behind technologically, even when this advice has doubtful long-term political implications. For instance, evidence that China or Pakistan is developing a new missile will tend to be exaggerated by these agencies or intentionally leaked to the press to whip up public concern about a "Chinese-Indian missile gap." In addition, now that India has missiles and atomic bombs, Delhi must face up to the problem of dispersing these in a crisis so that they do not draw fire, and to do this India must increasingly depend on its intelligence warning agency for advice.

It is not hard to imagine where this is heading. India has a rising computer software industry that is heavily funded by foreign multinational companies. India needs command and control to link these new forces, to pull them together into a coherent whole. Links between political leaders and the military are important to eliminate the potential of a surprise attack. The Indian intelligence establishment—the Research and Analysis Wing—must be brought into the picture because it is responsible for warning of attacks and assessing the strategic environment outside of India. It would be surprising if the future did not see horizontal communications links between all of these agencies.

Rajiv Gandhi's decision to build "small" atomic bombs has had consequences that transcend the short-term policy reasons that justified it. A more tightly interlinked Indian military-industrial complex is clearly in view, and this is as significant as any change in the armed forces since the creation of the Indian army in 1947. A civilian superstructure is growing up around the military services, offering political leaders a greater degree of control over the armed forces than they have ever had before. For almost the first time, India's unwieldy armed forces will actually be a useful tool in the government's foreign policy.

Strategic and technological decisions about weapons will increasingly have an impact on foreign policy decisions in ways unknown in the past. The cross-integration of the intelligence, military, and space programs is like nothing India has faced in the past fifty years. The circuits of political influence in the government will change, and business opportunism by industry to get more contracts is the wave of the future. Indian politicians, for example, will be faced with choices over entire technical systems, not just individual weapons. Whether to fund an antiballistic missile system, or how to build a deterrent that can survive a first strike, are decisions that necessitate new kinds of expertise. This at least has been the experience in other countries that have gone down the same road.

In India this sort of collaboration between government and the military has historically been resisted by political leaders. The first atomic bomb decision was made by Indira Gandhi alone, with only her very closest advisers in the know. The defense ministry was kept in the dark. Since independence India's rulers have rarely consulted anyone on questions of national strategy. But this will change as the technical component of India's security becomes more prominent.

There should be serious doubt about whether the new superstructures have the ability to perform effectively. India provides an example of what will afflict many other Asian countries in this respect. The three military services have never cooperated with one another, and each jealously protects its prerogatives and its share of the budget. At the same time, many of the weapons-building industries are less interested in national security than in booking new business and raising their status. Decisions about new weapons may well be made for essentially bureaucratic reasons, regardless of destabilizing international consequences.

For example, a decision to go ahead with a medium-range missile armed with atomic bombs could have the most serious consequences. It could provoke all kinds of counterreactions by Pakistan and China. Yet the decision could easily be made, with no consideration of the repercussions, by opportunistic bureaucrats who happen to have the ear of the prime minister. In the old system the army simply stymied civilian leaders with bureaucratic inertia. This was a stabilizing factor on the subcontinent, because it dampened military innovation. Its replacement by a more innovative system of scheming civilian technocrats bent on shoring up their budgets suggests a new factor driving Asian arms buildups.

NATIONALISM: THE "HOT" SIDE OF ASIAN DEFENSE

Giving atomic missiles to the Iranian Revolutionary Guards is not like deploying jets in the British air force. An irrational politics of rage permeates the Middle East, coloring the whole context there for missiles and weapons of mass destruction. This was not the case when Western forces first acquired nuclear arms.

Cold war passions were buried beneath an ideology of technocratic professionalism during the height of the Western arms race. The term *cold war* captured this point. It stayed "cold" because the competition was approached in a clinical, almost antiseptic way. The national hysteria that plunged Europe into hot wars was bottled up by the two superpowers. During the cold war the dangerous excesses of anticommunism and anticapitalism were checked when it came to military action. The professional and technocratic nature of the two military forces made this easier. There were no blood-curdling calls for revenge on the enemy from the American or Soviet officer corps. The cold war experience is worth noting in this regard. In both the United States and the Soviet Union it took about twenty years for the civilians in control to fully come to grips with the dangers of the nuclear age. In the United States President Truman's sacking of Douglas MacArthur clearly established who was in control. Later, in the 1960s, Secretary of Defense McNamara had to fire some generals to get this point across once again. In the 1950s the Soviet Union replaced its top generals with officers who were more technocratic than red, that is, more expert in managing complicated missile programs than heirs to the Bolshevik tradition of world revolution. The result in both the United States and the Soviet Union was that the political leaders on each side had little to worry about from hotheaded generals.

The transformation of Asia into a capitalist market economy is fostering nationalism, which reinforces tendencies toward extremism. While every Asian country is unique in its brand of nationalism, nearly all are increasingly exposed to the fragmenting forces of globalization and industrialization. Industrialization requires a money economy, which replaces the established social relations among classes. Peasants find it hard to adapt to the new conditions. Large-scale modern industry and commerce are especially disruptive to the small businesses and shopkeepers who are critical to the social fabric of most Asian countries. The power of big business grows, while that of small enterprise declines. Especially in countries like India and China, where custom governs so much of life, industrialization is likely to promote a breakdown in social cohesion as market mechanisms replace traditional forms of social control.

Something has to compensate for these shattering changes. It is no coincidence that nationalism is growing in India and China as they both undergo the effects of globalization. Nationalism finds fertile ground when old symbols and ways no longer provide the continuity they once did.

It is difficult to generalize about Asian nationalism, but one common thread connecting its many variations from one country to the next is an underlying anti-Western sentiment that goes back to the origin of these states. Asians from China to India, and throughout the Middle East, do not see themselves as living in a post–cold war era; instead, they see the West as trying to keep its grip on power through its domination of military technology and key economic institutions. Resentment is often below the surface and is tempered by other pragmatic considerations, but it is common throughout Asia. Nationalism harnesses all the immaturity and energy unleashed by the French Revolution and by communism in its expansionist heyday.

The new technologies of destruction must be seen against this background. Military modernization in Asia promises to be a more explosive process than the assimilation of missiles and nuclear weapons into Western forces in the 1950s and 1960s. The prospect that erratic organizations like the Revolutionary Guards in Iran will get their hands on modern weapons is, or should be, truly frightening to the West, because the structure of their military is so radically different from that of military institutions in the West. Most of these forces are not noted for their professional discipline. The Western framework of rational calculation of force balances—the number of missiles and bombs on each side—undergirds a cool professional attitude toward war that stands in contrast to the hot nationalist passions of Iran and other countries. The West responds to this dynamo of energy with the cold-blooded precision of its cruise missile strikes. But this way of seeing the military balance turns attention away from a more primal type of mass politics.

In the West a politics of distribution has taken hold: different interest groups vie for their share of the pie. The West no longer pursues, and therefore fails to grasp, an older politics of ideas, of fundamental clashes over ethnicity, religion, or the other touchstones of nationalism. In ideological politics, competition goes far beyond any rational calculation of costs and benefits. Wars over ideas are more intense than those for material gain. They leave no middle ground. This absence of a middle ground, of a split-the-difference mentality, justifies extraordinary losses, of a kind seen on the eastern front in World War II, and in Vietnam and Iran.

Nationalism is dangerously underrated by Western observers, who see it as part of a primitive political past that a nation sheds as economic progress leads to a more contented society. It is easy to forget that nationalism defeated Nazi Germany. Both the United States and the Soviet Union drew on emotional fervor, not abstract policy goals, to mobilize their tremendous war efforts against Hitler. In the 1960s Vietnamese nationalism defeated the strongest military power in the world, the United States. It legitimized extraordinary losses, losses that Washington planners never dreamed North Vietnam could bear. Nationalism drove the United States from its strategic position in Iran in 1979; it mobilized Iranian youth to charge machine-gun fire in the war with Iraq in a display of bravery and folly the world had not seen since 1917.

The most important issue of the twenty-first century is understanding how nationalism combines with the newly destructive technologies appearing in Asia. If the energies in these examples become harnessed to this firepower, then the world could be in for a very dangerous era that the West is ill prepared to understand. Thus, the principal dangers in an Asian arms race are not the ones usually imagined in Western analyses, with their rational calculation of deterrence. Balancing the size of arsenals, promoting stability, mutual confidence, and openness—these are Western, technocratic approaches to international security that may not have that much to do with the intensities that animate rivalries in Asia. Concerns over miscommunications in a crisis evoke a universal solution in the Western mind: the hot line, the red telephone link between two heads of state that is used only when the chips are down. But Asian nations don't see disputes as always arising from miscommunications. There is a hot line connecting India and Pakistan, intended to dampen their dispute in Kashmir. The problem is that it is used all the time, and thus both sides ignore it. It has none of the drama that the Washington-Moscow hot line had during the cold war.

No one is against good communication. But the sources of instability in Asia are ones that cannot be eliminated through hot lines and high-tech locking devices to prevent the unauthorized launch of weapons. It may be better to have these safety measures in place than not to have them, but they tend to divert attention from the more primitive animosity that lies below the surface and can be inflamed by opportunistic politicians. The mass politicization of military competition is a primary source of instability and danger in Asia. What puts arms races into overdrive is the appearance of falling behind, which creates an overwhelming impulse to catch up, even if there is no catching up to do. Imaginary "missile gaps," created and exaggerated in the Asian mass media, resonate among nations with ancient histories of mutual hatred and long-standing feelings of inferiority. The frenzied response in the United States over missile and bomber "gaps" in the cold war would seem tame compared to some possible missile hysteria in Asia.

Just as having an army was the main symbol of being a serious nation-state in the postcolonial period, now bombs and missiles are the new symbols of modernity. China tested a nuclear weapon in 1964, and India matched it in 1974. These technologies occasion new avenues for strategy and raw self-interest. They enlarge the strategy space for new kinds of interlinkages. North Korea is a good example. North Korea's programs in nuclear, biological, and chemical arms, intended to offset South Korean and American military strength, began in the 1980s. In the 1990s the North faced another kind of pressure, its own economic disintegration with the prospect of internal collapse from starvation.

North Korea has now linked its national survival to that of its neighbors through a dangerous but nonetheless carefully designed military structure built on weapons of mass destruction and missiles. Its survival presses sharply on that of its neighbors because for Pyongyang weapons of mass destruction have become a way to link these internal and external pressures. Outsiders had better not pressure North Korea to implode, or they might cause it to explode. Since North Korea is well aware that any invasion of the South would be suicidal, the threat takes this into account. In effect, Pyongyang threatens to commit suicide by blowing out its brains all over Northeast Asia's living room.

North Korea's famine and economic failures are thereby "exported" to the world. Outside powers would much rather ignore the North, but they must deal with the missiles and atomic bombs it builds, supplying concessionary food, oil, and medicines even as the country continues to put most of its effort into weapons of mass destruction.

Just as internal traumas can be exported, both real and imagined evils can be imported, to distract attention away from the government's failures. India can "find" threats in the Indian Ocean with its extended military forces. Chinese naval power projection past the Straits of Molucca is a gift to opportunistic Indian politicians, for it provides an occasion to counter India's centrifugal national politics. Indian nationalism is an especially interesting case: India's fragility as a coherent state requires extraordinary measures to keep it unified around something. The Nehru-Gandhi dynasty through the Congress Party established a secular nationalism. They felt that India was too diverse to ever unite around any national identity based on religion, caste, or language. Instead, they espoused modernization as a way to pull India together. This was an effort to unite India around a belief in progress, economic development, secularism, tolerance, and the power of the central state. Over the years this effort gradually stalled as Indian politics became more fragmented and chaotic. Fundamentalist parties and sentiment grew in India, with a corresponding loss of momentum in the Congress Party's secular nationalism.

The nuclear dimensions of this change are important. The first Indian bomb was designed to reinforce secular nationalism, showing that India was a modern power capable of keeping up with the Chinese. It was a statement of what secular India could do. But with the decline of secular nationalism, the context of India's bomb changed, too. When the fundamentalist Bharatiya Janata Party (BJP) was elected in 1998, it seized on testing the bomb to consolidate its power, based on different sources of nationalism. Although nearly all of the second bomb program originated under the Congress Party, the BJP recognized an opportunity when it saw one. By ordering the nuclear tests, and heating up the war in Kashmir, the BJP appealed to India's new, darker form of nationalism to gain support for its regime.

China, likewise, can heat up the Taiwan issue at any time if its leaders sense they are losing their tight hold on power. The difference now is that India and China have the military reach to do something beyond railing about their grievances. Thus, the temptation to take action can back governments into a corner more easily than in the days when armies were slow to mobilize and limited in reach.

The political leaders will always believe themselves in control of the situation. But they can be deluded. Sensationalistic distortions or charges of "falling behind" or "betrayal" could whip up public hysteria, raising tensions throughout whole regions. What begins as bald political opportunism—like the BJP testing five A-bombs—could in the future be reinforced from below by street politics that fracture the government's control over events. Along with the Asian military-industrial complex, a disorganized, explosive new mass politics is straining the limited control of Asia's old regimes.

Finally, there is the most basic point of all. Nationalism has made a second nuclear age. The impetus for missiles and weapons of mass destruction in the arc of terror that extends from Israel to North Korea is *national* security. The question of resources, of whether nations are spending too much on defense, has never been a serious constraint because it is national survival at stake, not the budget politics that predominate in Western defense decisions. Even the small impoverished countries have come up with the money to join this club.

4   THE SECOND NUCLEAR AGE

The Indian atomic tests of 1998, quickly matched by Pakistan, provoked the expected reaction from the United States, which views itself as the conscience of the world when it comes to nuclear nonproliferation. Washington invoked global sanctions against the spread of weapons of mass destruction. It pressed New Delhi and Islamabad to sign a comprehensive test ban treaty and urged both not to arm their missiles with atomic warheads.

At the same time, though, Washington had to acknowledge that nuclear weapons had come back from the realm of the presumed dead. Their disappearance from world politics was widely predicted after the end of the cold war. Nuclear weapons were marginalized, thrust from the center stage of international relations to the hinterland of a few oddball think tanks. Most experts in academia and the government hoped this was a step along the road to their complete abolition, turning back the clock to the world before 1945.

India and Pakistan had broken free of Western nuclear controls, and in the late 1990s other countries in the arc of terror followed suit. The breakout was legal and technical. Legally, the Western arms control effort that had worked for twenty-five years was violated time and again. Every country in the arc of terror, from Israel to North Korea, pushed the envelope. Most got away with it. Even Iraq would have succeeded if it hadn't made the mistake of attacking Kuwait, opening itself to intrusive inspections.

Technically, the emergence of a wide variety of disruptive technologies overwhelmed the West's capacities to monitor and control them. Keeping track of missile launches, underground bomb plants, and chemical and biological warfare programs in various countries led to proliferation fatigue. So much political capital was invested in containing Iraq that little was available to take on the tougher cases of North Korea and Iran, let alone those with white-hot political sensitivities, like China and Israel.

The world is in a second nuclear age, an Asian nuclear age. It is a second age because it has nothing to do with the central fact of the first nuclear age, the cold war. For China, Israel, India, and Pakistan, the cold war is an antiquated irrelevance. It was only in the notoriously provincial West that the collapse of international communism was seen as the great dividing line of history.

It is easy to forget just how much the cold war was a European affair that spilled over to other regions, and how deeply Eurocentric nearly everything about the first nuclear age really was. The bomb was built on principles discovered by European physicists. Fears of a Nazi A-bomb drove the U.S. Manhattan Project. Thus, the first nuclear age was an outgrowth of the two great European wars of the twentieth century. Deep psychoses of violence within European civilization were at work, bringing forth the "isms"—communism, Nazism, fascism. These thoroughly European ideologies produced poison gas, Dresden, Auschwitz, and the doctrine of "total war." Hiroshima and the cold war did not come from a historical vacuum but from identifiable European origins. The great civilizations of China, India, and the Middle East understand this and accordingly reject the idea that the European states are the only ones that can be trusted with the awesome responsibility of the A-bomb.

The Asian states have learned from the West. They have learned how to use nuclear weapons without actually detonating them in an attack, for political maneuvers, implicit threats, deterrence, signaling, drawing lines in the sand, and other forms of psychological advantage. The United States now forgets how it "used" nuclear weapons for forty years to reshape international politics to its advantage—when it deterred Soviet conventional attack on Europe, for instance, or when it went on high alert during the Cuban Missile Crisis and the 1973 Middle East War. The same pattern is appearing in Asia. North Korea uses its (implied) nuclear weapons to thwart outside pressure for reform and to extort free food and oil from the West. India uses nuclear weapons to send a wake-up signal to the United States about its relationship with China, its refusal to accept second-class status among the world's powers. Israel uses nuclear weapons to intimidate the Arabs and to play on their subconscious sense of inferiority. Whether these uses are successful or prudent is beside the point. The American position, on which a quarter-century of foreign policy has been based, is that there are no legitimate uses for nuclear weapons, and therefore no reason for additional countries to aspire to them. This turns out to be not quite the case, and when Washington finally wakes up to the fact, it will come as a rude surprise.

THE FIRST NUCLEAR AGE

The first nuclear age started on July 16, 1945, with the testing of the American atomic bomb in the New Mexico desert; Hiroshima and Nagasaki were attacked three weeks later. It was their bad luck that Germany, the original target of the Manhattan Project, surrendered before the bomb was ready. But the history of the nuclear age is filled with such revisions of thinking and plans. Some of the lessons of the first nuclear age are useful to review in seeking a better understanding of some of the dynamics, and pathologies, that could shape the second.

"There Are Atoms in All Countries"

Almost no one, at the start of the first nuclear age, expected that atomic weapons would remain a Western monopoly for long, and many people believed they would be used in warfare again eventually, certainly before the end of the century. "There are," Albert Einstein famously remarked, "atoms in all countries." Few people believed that international cooperation would stop the spread of atomic weapons; President Kennedy thought that by 1980 "everybody" would have the bomb—the Germans, the Japanese, the Brazilians, even the Swiss. Herman Kahn, by no means the most pessimistic voice, in 1960 predicted that by 1975 there would be between five and twenty nuclear powers. His low estimate proved nearly dead on the mark, for by 1975 only six countries had the bomb—the United States, the Soviet Union, Britain, France, China, and India. More remarkably, by 1995, twenty years past Kahn's forecast, no additional countries had joined the club. The pessimists were wrong because arms control turned out to be much more effective than anyone thought it would be. It worked by both formal treaties and, just as important, informal but widely observed norms of international behavior.

Arms control had been a feature of great power relations since the nineteenth century, but the terrifying consequences of atomic war lent it a new importance. Public sentiment in Western Europe and the United States demanded it, while even in the Soviet Union arms control sentiments gradually took hold in political, if not military, quarters.

In particular, the 1968 Nuclear Non-Proliferation Treaty (NPT) worked far beyond anyone's expectation of success. Originally it was seen as a way to buy time, maybe five years, until a more permanent structure came into being, but it lasted much longer, ostracizing countries that tried to get their hands on these weapons. These pressures stopped the bomb in Brazil, Argentina, South Korea, and Taiwan. South Africa even went from being a nuclear to a non-nuclear weapons state, dismantling its bombs following the end of apartheid. International condemnation forced those who wanted the bomb to go underground. Desiring an atomic bomb—for those countries that didn't already have one—became viewed as a kind of perversion, something countries did, if at all, in deepest secrecy. The result was that not many bombs were built.

The Surprising Strength of Deterrence

If the greatest surprise of the first nuclear age was that more countries didn't build nuclear weapons, the second biggest surprise was how well deterrence worked. This was not at all apparent when nuclear weapons were first deployed in large numbers beginning in the 1950s. But it turned out that deterrence was surprisingly resilient in the face of the shocks of the cold war. Through numerous crises and regime changes, the one constant was that the missiles stayed in their silos. The world owes more than it realizes, perhaps, to the leaders of the two nations whose cautious political maneuverings over forty years avoided confrontations that could have led to nuclear war.

Unfortunately, this has led to a widespread belief that as soon as two antagonistic countries get the bomb, deterrence is "automatic." If everyone got the bomb, by this thinking, the world would be a much safer place.

But the history of the first nuclear age suggests that things are not quite so simple. In effect, there were two cold wars. The period from 1947 to 1967 was much more dangerous than the years from 1968 to 1991. Virtually all of the nuclear close calls occurred before 1967. Showdowns between Washington and Moscow over Berlin, Panmunjom, Taiwan, Beirut, and Cuba were all serious enough for American field commanders to ask the White House for permission to ready atomic weapons. All of these took place in the first twenty years of the cold war, the cold-war learning curve. During this time both sides felt each other out to see how they could use nuclear weapons to their advantage. They were learning, gingerly, the tactics of "brinkmanship," of maneuvering for advantage on the edge of an abyss. Each side tried to use the danger of falling over the brink to get its way, without actually fighting. This was a reasonably accurate metaphor for the early nuclear age.

In the 1947–67 period both sides also turned over thousands of nuclear weapons to their armed forces, loosening the political tethers of control. At the same time, "fail-safe" systems were installed so that false alarms did not lead to automatic launch with irrevocable consequences. Layers of controls were added to prevent the hypothesized "mad major" from starting World War III by himself.

These systems did their job, but by the latter part of the cold war, another factor came into play, as each side built forces that could not be easily located and destroyed by the enemy. Missiles buried in underground concrete silos or aboard submarines at sea were invulnerable to a surprise first strike, reducing the need to stay on permanent, hair-trigger alert. "Second-strike" arsenals acted like huge shock absorbers for the clashes of the second half of the cold war. For all the "Evil Empire" rhetoric of the Reagan administration, in a military sense the confrontations of those years were nowhere near as dangerous as those of the 1950s.

But the world had to live through twenty perilous years before it reached this stage. The arsenals became as safe as they did only because of huge efforts by the U.S. and Soviet militaries. It did not happen automatically, and it cost trillions of dollars. Much of this astounding expenditure went not for the nuclear buying spree itself, but for training and command and control systems to make the weapons that were bought safe from inadvertent use, or from necessitating destabilizing first-strike strategies.

The "Use" of Nuclear Weapons

Although the overall trend of the cold war was toward greater moderation and safety, there were still several very close calls. There were two major wars in Asia, showdowns over Berlin and the Middle East, and the perilous Cuban Missile Crisis. In each of these events, and many others, nuclear weapons were used by the superpowers. They just weren't fired, at least not at each other. The United States moved a nuclear cannon into Korea in 1953 after publicly releasing movies of this weapon firing atomic shells into the Nevada desert. In Europe the United States matched the larger Soviet army with thousands of "small" tactical nuclear weapons. For NATO, nuclear deterrence was the principal counter to Soviet intimidation. During the Cuban Missile Crisis in 1962 President Kennedy launched B-52 bombers loaded with hydrogen bombs, manipulating the risk of nuclear war to get the Soviet missiles out of Cuba.

Both parties used nuclear weapons to send political signals, sometimes publicly, sometimes covertly. When Nikita Khrushchev wanted to show his displeasure over American intransigence on Berlin in 1958 (the United States would not recognize that the Soviets were giving the East Germans control of the eastern sector of the city), he communicated his unhappiness in a unique way: he ordered the detonation of an enormous fifty-eight-megaton hydrogen bomb high over the Arctic. In the process, he broke the world's first nuclear test ban agreement between the two sides, an informal agreement reached in the preceding months to lessen tensions between them. His idea was to break the agreement as a way of showing just how annoyed he was over Berlin.

When the United States wanted Moscow to talk Hanoi into accepting the Paris peace accords to end the Vietnam War, it prepared its worldwide nuclear forces for military strikes. B-52s were brought back from Europe and Guam and refitted at nuclear bases, creating the strong impression in Moscow that they were being fitted with atomic bombs. This was all part of President Nixon's so-called madman theory—intentionally creating the impression that Nixon would do almost anything to break the diplomatic deadlock in Vietnam, including issuing nuclear threats. Once again, nuclear weapons were "used" for political purpose.

At other times, nuclear weapons were used to rattle the other side's nerves as a general matter of principle rather than in connection with any specific negotiation or deadlock. Part of President Ronald Reagan's cold war strategy was to intimidate the Soviets by operating submarines, aircraft carriers, and jets conspicuously near Soviet borders. The Soviets did the same to the United States. In 1971 they began deploying nuclear-armed submarines off the East Coast of the United States, where they could hit Washington in just eight minutes. In the 1980s the Soviets reacted to the Reagan administration's stepped-up defense programs in a similar way. Moscow undertook near-simultaneous tests of new ICBMs, submarine-launched ballistic missiles, and even an antisatellite weapon. In the span of twenty-four hours, all of these were shot off, lighting up the warning screens of the CIA. Never a word was spoken between the two sides about the tests. Rather, the United States was simply left to figure out what might come next if it continued on its buildup path.

Apart from the great strategic showdowns, nuclear weapons provided the backdrop to the grunt work of the cold war, the routine harassment and "nibbling" that occupied so much energy for so long:

in Poland and Afghanistan, where the United States helped fuel political and military resistance to Soviet rule, and in Central America, Africa, and Vietnam, where Moscow did its best to keep Americans running in circles. The trick was to put the burden of escalation on the other side. None of these conflicts was, by itself, worth fighting a superpower war over; ironically, having nuclear weapons probably encouraged these low-level torments, precisely by ensuring that Americans and Russians would stop just short of shooting at each other.

Strategy and Nuclear Weapons

The evolution of nuclear strategy in the first nuclear age was complicated, not because it's such an arcane subject, but because declared strategy was not the same as actual strategy, and what passed for strategic debate among specialists often had little or nothing to do with the bureaucratic momentum behind real plans.

The very first impulse was to declare that nuclear weapons made strategy obsolete. To use them was regarded as tantamount to suicide, so there could be no nuclear strategy. Bernard Brodie, one of the early civilian nuclear strategists, argued this point in a famous article titled "Strategy Hits a Dead End." The logic of his essay seemed sound enough, but Brodie did not delve into the bureaucratic momentum of the arms race; this oversight contained the undoing of his argument. As large numbers of weapons were churned out in the 1950s, they had to be aimed at something. Brodie's abstract arguments could not eliminate the necessity this created as the military put together targeting plans for atomic war. A strange juxtaposition (certainly not the last of the nuclear age) resulted: just when arguing that there was no such thing as a nuclear strategy became a mark of intellectual seriousness in some circles, plans for just such a strategy were being drawn up by the U.S. Air Force, and by the Soviet Strategic Rocket Forces as well.

But there was a gap between declaring a strategy and the ability actually to implement one. The gap between the public posture and the military reality of what could be done was often enormous. For example, in the 1970s the White House ordered up a strategic plan for a "controlled" nuclear war; that is, a war that would stop short of complete annihilation with direct attacks against population centers. The wisdom of this was hotly debated. But in fact, neither the United States nor the Soviet Union could have executed such a strategy; their command and control structures weren't up to it. It is useful to remember this in thinking about the new nuclear states. Many of their academics and officials attend Western conferences where they pick up the language of nuclear strategy and arms control. But there may not be much relationship between what such experts say at conferences and what their governments plan and do in secret.

From Pearl Harbor to Desert Storm

Armies, it is said, are always preparing to fight the last war. The West bravely entered the nuclear age with a stock of metaphors and intellectual models drawn from its experience in World War II: of nuclear Pearl Harbors and Munichs, of an atomic-powered Hitler blackmailing the world. These images influenced strategic thinking for decades, but they have been replaced for a new generation of foreign policy specialists by the Cuban Missile Crisis, Vietnam, and the Gulf War. In the United States especially, these three metaphors now seem to exhaust all historical experience. The Cuban Missile Crisis has been studied to death, and we apply at our peril its "lessons" to future nuclear showdowns that may in fact call for very different responses. Vietnam, in contrast, is studied as a caricatured model for disastrous decision making. And the Gulf War is seen as the exemplary triumph of American technology. Each of these newer metaphors contains dangerous simplifications, just as their

Pearl Harbor and Munich predecessors did. Yet they are the ones most likely to be applied by policy makers in the future. Western strategists have yet to make the leap beyond their traditional metaphors, ones that have very questionable utility when applied to Asia.

The Illogic of Arms Races

Fear of nuclear war didn't stop the two superpowers from building thousands of nuclear weapons. Such arms races occurred in spurts all during the cold war. Their rationale and timing defy logical explanation. Once the two sides understood the mechanism of deterrence, there would appear to have been little reason to keep piling up additional weapons. But that is exactly what happened: just as deterrence stabilized in the late 1960s, each side began a huge building program. The number of nuclear weapons in 1991 was far greater than in 1967.

This suggests that motivations were bureaucratic and political, rather than strategic. The cold war was characterized by sharp and seemingly inexplicable shifts in intensity. In the late 1940s the Soviet Union installed puppet governments in Eastern Europe, broke its World War II agreements, and positioned the Red Army within striking distance of the Rhine. But the United States hardly reacted. Defense budgets stayed low, and only a dozen nuclear weapons were built as all of this happened.

At other times, though, relatively small provocations triggered a large American reaction. Moscow's deployment of new medium-range missiles in the late 1970s—missiles that had little overall impact on the balance of threat in Europe—gave rise to a great surge of indignation that paved the way for the large defense budget increases of the Reagan years.

The lesson, which is likely to be demonstrated again in Asia in the years to come, is that rational military and strategic calculations don't always apply in the realm of nuclear weapons. Accidents of timing, of the order in which things occur or the proximity of domestic elections, can affect events in ways almost impossible to predict, or even to understand in hindsight. What passes for a principle often turns out not to make much sense even a few years later.

The defense of Europe—in which, after all, lie the origins of the cold war—provides a good example of this. In the 1950s the United States built strategic nuclear weapons because the Soviets had outmanned it in that area. The "nuclear umbrella" substituted for conventional armies, allowing the United States to keep a relatively small force in Europe. But over the decade, as the Soviet Union's arsenal grew, Europeans began to question whether the United States would really risk the destruction of New York and Washington to save Bonn and Paris from a Soviet invasion. A new strategic doctrine therefore called for beefing up American forces in Europe, not so much to match the Red Army's infantry and tank strength—an impossible goal—but to ensure that any Soviet invasion would entangle the United States to the point where it would have no choice but to defend its allies. This was a 180-degree reversal of the earlier strategy. In 1958 nuclear weapons obviated the need for a strong army. In 1962 they made the army more important.

The Europe case is actually more bizarre than this. When the NATO allies balked at accepting the new emphasis on conventional forces, Washington had to compensate by quadrupling the number of tactical nuclear weapons deployed there. Thus, the era of conventional defense of Europe was actually a period of sharp nuclear buildup, the very opposite of the stated purpose of the new policy.

At times American strategy was equally unclear to its allies, its enemies, and the American public. In the late 1970s President Carter lobbied hard with the NATO allies to deploy the neutron bomb, a tactical weapon against Soviet tanks. It was extremely unpopular among Europeans, but the issue was cast as a test of NATO's will and its capacity for modernization. Blocking it was thought to signal the crippling of NATO. Finally, the allies grudgingly accepted the neutron bomb, at great domestic political cost. Then, for no apparent reason, the president abruptly canceled it, as if the earlier strategic justifications no longer mattered.

The Great-Man Theory

It might seem that the nuclear arms race was driven just by its own deadly logic. But the first nuclear age was shaped by the leaders in power at the time—Stalin and Eisenhower, of course, but also those who came after, like De Gaulle, Kennedy, and Khrushchev. The French nuclear program was adrift in the 1950s and took on a strong direction only when Charles De Gaulle became president in 1959. John F. Kennedy greatly expanded the U.S. nuclear arsenal. He came in person to observe test-firings of new missiles, part of a campaign to publicize the dynamism of his administration. Kennedy's goal was "to get America moving again," and he took every opportunity to contrast his administration's energy with the lethargy that he attributed to his predecessor. The most intense nuclear crises, in Berlin and Cuba, took place during his administration, and these cannot be understood without considering his personality—or that of his adversary, Nikita Khrushchev. The flamboyant Khrushchev was quick to make decisions without thinking through all of their consequences. The world was thrown into its most intense nuclear crisis, in Cuba in 1962, because of the interaction of these two leaders. It should not be forgotten how much of the first nuclear age depended on the unpredictability of the personalities of the leaders who happened to be in charge, and on the way they sometimes interacted. These factors often counted for much more than the numbers of bombers and missiles in the arsenals.

In contrast, Khrushchev's successor, Leonid Brezhnev, avoided showy displays. His personality gave rise to a methodical nuclear buildup that was immune to any U.S. arms control proposal or even to any counterprogram. The Soviet Union under Brezhnev seemed autistic. No matter what the United States did, the Soviets built more weapons.

Leadership made a big difference in the first nuclear age. And changes in leadership often meant fundamental shifts in strategy and weapons.

The Will to Bomb

Finally, one important lesson of the first nuclear age was that while the military capabilities of nuclear forces mattered, the will to use them mattered more. No amount of spending could achieve a tenacity of purpose, and credibility in using force was much more important than numbers of weapons.

Thus, Britain was never taken seriously as a nuclear power even though early on it had a fairly impressive arsenal. Through the 1950s London could have destroyed the top ten Soviet cities with its Vulcan and Canberra nuclear bombers and could have done more damage to the Soviet Union than Moscow could have done in turn to the United States. But Britain was simply not taken seriously as a major power by anyone, not even the United States.

China, in contrast, had few bombs and an army it could barely move. But few doubted China's will, and it was taken very seriously. Any objective comparison between China and the United States showed that Washington far outstripped Beijing in every measure of power. But the United States was scared of China and even considered ways to keep it from becoming a nuclear power. When Beijing went nuclear in 1964, the Johnson administration drew up serious plans to take out its atomic capabilities. China was the decisive factor in U.S. strategy in Vietnam. Washington limited attacks and was willing to accept a defeat with 50,000 killed rather than risk drawing China into the war.

THE SECOND NUCLEAR AGE

The second nuclear age is hard to date precisely. Arguably, China's 1964 test of the first Asian atomic weapon marks its beginning. For the second time an Asian country acquired advanced weapons that only the West had possessed. But at the time the West viewed the Chinese test from the perspective of the cold war and the superpower confrontation. No one related it to the modernization of Asian forces that had begun in Japan early in the century.

What the Chinese test did accomplish was to make the whole world nervous about nuclear proliferation. After painful reflection wherein both Washington and Moscow alternately considered a decapitating strike against China, Beijing was allowed into the nuclear club, with the door slammed shut behind it. The Nuclear Non-Proliferation Treaty allowed only five nuclear powers, the United States, the Soviet Union, Britain, France, and China. (In fact, China did not sign the treaty until 1991.) There was no provision for any other countries to build or obtain nuclear weapons. New NPT signatories could join only as non-nuclear powers, a restriction that made arms control look like a club whose membership rules were controlled by the West. The only purpose of China's grudging acceptance was to draw the line around any further expansion of the club.

So perhaps the second nuclear age really began in 1974, with the first Indian shot, because theirs was the first outlaw bomb. India built its bomb outside the legalistic Western arms control system. Even after exerting great pressure, the West couldn't persuade India to budge—to give up its bomb, accept inspection, and join the NPT as a non-nuclear state. The Indian test meant that the West's attempt to draw the line against further spread of the bomb had failed.

Again, maybe the second nuclear age began between the Chinese and Indian shots, when Israel developed its first bomb; never officially tested, it probably was built around 1969. Israel had also refused to sign the NPT, and its bomb was also built outside of the cold war, even before India's.

Because it is hard to say exactly when the second nuclear age began, many people aren't fully aware that it wrought great changes, all of which occurred gradually, obscured by a cold war that had very little to do with it. Because nuclear weapons spread so slowly, much more slowly than pessimists had predicted, it was possible to overlook the fact that they were spreading at all. The second nuclear age seemed to emerge out of a hodgepodge of unrelated regional issues: Chinese desire for status, politics on the subcontinent, Middle Eastern security. The larger connection was missed.

In a world attuned to rapid change, slow-motion change does not register on most scales. There was no news bulletin announcing the start of the second nuclear age—as there was, of course, when the first nuclear age became public at Hiroshima. The ease with which the unhurried spread of the bomb could be overlooked made that spread similar to many other changes in the world. In the decades before World War I Europe was peaceful, and international relations were carried on much as they had been since the middle of the nineteenth century. Statesmen saw no need to do anything differently simply because new weapons like the machine gun, the submarine, the airplane, and poison gas had come about. Then came 1914, and the world changed forever in a few months.

The spread of the bomb to Asia makes the world less Eurocentric. The second nuclear age will greatly accelerate the globalization process already under way. Now Asia is generating events on the world stage, not just reflecting the interplay of forces in Europe and the cold war. If Asia is not viewed that way, on its own terms, we will be as surprised by the next great conflict as the diplomats of Paris and Berlin were by the outbreak of World War I.

Nationalism and the Bomb

The greatest single difference between the first and second nuclear ages is nationalism. The cold war was more of an ideological struggle. For the Soviet Union, consisting of many disparate nationalities held together mostly by terror, nationalism was a dangerous force. American nationalism, for its part, was tempered by Washington's role as leader of "the free world," an anti-Communist alliance embracing nations as disparate as Britain and the Philippines (or Greece and Turkey).

The link to nationalism makes the second nuclear age even harder for the West to comprehend. Nationalism is not viewed kindly in the West these days. It is seen as nonsensical, a throwback, and, it is hoped, a dying force in the world. The notion that the Chinese or Indians could conduct foreign policy on the assumption of their own national superiority goes against nearly every important trend in American and West European political thought.

The second nuclear age is driven by national insecurities that are not comprehensible to outsiders whose security is not endangered. Its metaphors are fundamentally different from those of the cold war, grounded in Munich and the Cuban Missile Crisis. Each state in Asia sees the world in its own terms, drawn from its unique history and situation—and frames its nuclear ambitions accordingly.

For Israel, the bomb links the deepest fears of the Jewish people to the ultimate form of modern destruction. Whatever actual military plans may call for, the Holocaust and Masada are metaphors that define the scenario for nuclear weapons in the national consciousness.

The Chinese, by contrast, have a legacy of greatness stretching back to antiquity, and a more recent history of oppression at the hands of Westerners. For them, possessing the bomb is a natural way to reclaim their past. The metaphor is one of the bomb righting past humiliations. China was treated badly by the West, to be sure. But China shares responsibility, having succumbed in part to its own corruption—as Japan and Vietnam did not. The chip on its shoulder is partly of its own making.

In India and Pakistan nuclear tests set off public euphoria—literally, people danced in the streets. This emotional embrace of a technology Westerners have been taught to loathe and abhor was probably more shocking to American officials than the explosions themselves. It is an outgrowth of the fierce nationalism that animates these two countries and the ethnic and religious rivalries that have defined them since independence—forces that proved far more powerful than the arms control machinery of treaties and sanctions and the weight of world opinion.

Weapons of Mass Destruction

Even at the worst moments of the cold war, at least most people didn't lie awake nights worrying about anthrax bombs or nerve gas. The United States and the Soviet Union, it is now known, did produce chemical and biological weapons, but they were never integrated into the forces or strategy of either side. In the United States biological weapons were looked down on by the nuclear scientists as somewhat immoral, and the important weapons labs had nothing to do with them. The Soviet Union's immense stocks of these weapons mostly stayed in warehouses.

The situation is different in the second nuclear age. Iraq has already used chemical weapons against the Kurds, and it was prepared to use them at the end of the Gulf War. The North Koreans have a vast program that thoroughly integrates chemical weapons into its army. Cultural prohibitions against them do not seem to carry the weight they do in the West. They are considered the poor man's atom bomb, a way to compete with more advanced enemies. Arab states, for example, obtained chemical weapons in the early 1970s following Israel's rapid victory in the 1967 war.

Nationalism among the new nuclear states in Asia makes for a different approach to strategy. *Dr. Strangelove* notwithstanding, in fact the United States and the Soviet Union approached nuclear warfare with detachment and rationality. Game theory and systems analysis were used to arrive at cost-benefit killing ratios and levels of unacceptable damage. In war games, the West took its worst nightmare, Hitler, and gave him nuclear weapons. Strategy was approached like poker and chess, with bluffing and penultimate showdowns. The Cuban Missile Crisis, the classic case of cool decision making under pressure, became the leading metaphor for cold war crisis management. Strategy became the purview not of the generals but of outside think tank experts. It is hard to imagine such a development in Asia. The idea of budding defense intellectuals sitting around with computer models and debating strategy in Iran or Pakistan defies credulity.

But that's just where Western models fall down. Deterrence under conditions of emotionally charged mass politics—hysterical nationalism—in which rage and religious hatreds drive decisions is something the Western think tank analysts never considered. That is one of the crucial differences in the second nuclear age.

Buy the Bomb!

Another significant fact about the new nuclear powers is that they are almost all poor by Western standards. All of the first nuclear powers were wealthy, industrialized countries, with the exception of

China. Even the Soviet Union was a major industrial power. None, certainly, had huge peasant populations in need of the bare essentials of life. For decades it has been argued that the need for schools, health care, and roads would take priority in Asia over the desire to buy atomic bombs. But this argument no longer holds; North Korea and Pakistan are among the poorest countries in the world.

Instead, the poor countries are amortizing the cost of the bomb and financing new weapons projects by exporting military technology to even less developed countries. The main response of North Korea and Pakistan to Western economic sanctions has been to step up their foreign sales of weapons. No one approves of this, of course, but their economic position is so desperate that it cannot be all that surprising.

In the first nuclear age, sharing military technology was based on strategic, not business, factors. The United States helped Britain, and the Soviet Union helped China. Neither did so for the money. But the more pressed Asian countries have been by the recent financial crisis, the more willing they have become to sell even the most sensitive and dangerous technologies. North Korea and Pakistan have nothing else to export.

OBJECTIVE PROBLEMS OF WEAPONS OF MASS DESTRUCTION IN ASIA

Even a gradual spread of weapons of mass destruction is likely to intensify certain problems in Asia. So far, so much attention and energy has been given to preventing the spread of weapons of mass destruction that there is relatively little objective thought given to the problems that such a world engenders. This focus is dangerous, much like refusing to think about the problems of the first nuclear age.

Arms Races

Arms races in Asia do not approach the scale of the cold war, and it is hard to see such a scale ever being reproduced there because the countries do not have the money for it. But low-level arms races could still occur and they could erupt into major confrontations. Asian arms races have received very little attention, and some of the obvious dangers are overlooked.

Israel and India are both experimenting with submarines to fire nuclear missiles. This shows a recognition of the vulnerability of their fixed missiles to enemy first strike. During the 1973 Middle East War, Israel moved some of its nuclear weapons out of their storage bunkers in Dimona to prevent them from being struck and to prepare for their use if needed. During the Gulf War, Iraq fired several SCUDs at Dimona in an attempt to destroy the weapons stored there and the Dimona nuclear reactor, a hit that would have spread radioactive contamination over a wide area.

Yet arms races in Asia might take a form very different from those of the cold war. China, for example, has no need to take on the United States in strategic nuclear forces. It only has to be strong enough to threaten vulnerable U.S. bases in Asia, while maintaining a minimum nuclear deterrent against the United States. It is hard to imagine the United States ever risking a nuclear confrontation with China, so Beijing needs only a "small" force to deter Washington from taking risks over Taiwan or other trouble spots. By dominating the Asia-Pacific periphery with short- and medium-range ballistic missiles, China seizes the advantage. These weapons put the burden on the United States of moving forces forward and into harm's way.

China is building large numbers of these short- and medium-range missiles but has only a small number of ICBMs that can reach the United States. This buildup focuses on the lower rungs of the escalation ladder and makes China a more formidable regional power, while avoiding a global confrontation with the United States. Still, Chinese espionage against American nuclear weapon labs, and the subsequent construction and testing of bombs based on these American designs, shows that China might well increase the ICBM threat to the United States homeland itself. With the new design, many smaller lightweight warheads can be packed on a single missile. This gives the Chinese ICBM force a tremendous new punch at bargain-basement prices because the United States paid for all of the research and development. The strategic intent of the Chinese ICBM forces is likely to focus on deterring the United States from taking military and political actions in Asia, in essence, restricting Washington's maneuvering room on the chessboard.

Asian arms races might take on a form that was also seen in the cold war—building weapons with the intention of driving an opponent into bankruptcy or political turmoil. The poverty in parts of Asia could make this an especially devastating tactic. India could easily force Pakistan into a desperate economic situation, for the simple reason that Pakistan is nearly there already. Indian superiority in engineering, nuclear science, computers, and software is likely to continue. India could build a few prototype missiles, and then publicize them, to drive Pakistan into even more wasteful arms expenditures. Were India simply to announce a development program of tactical nuclear weapons—without even going ahead with the program—the stress on Pakistan would be enormous. If Pakistan tried to catch up, as seems likely, the crippling economic cost could easily lead to political disorders. It is much more likely that India would use its nuclear weapons against Pakistan this way than by actually waging a war.

Shaky Command and Control

Command and control problems could be major sources of instability between Asian countries. Early warning of attack in Asia still depends greatly on spies placed in the enemy country, as in the earliest part of the cold war when the CIA and the KGB had to give warning based on their spy networks, before U-2 airplanes and spy satellites were available to provide a better basis for warning of attack. Satellites and reconnaissance airplanes are showing up in Asia, but coverage is spotty, and warning still depends on the spy in the enemy's headquarters.

Spy reports, however, are notoriously unreliable. In crises between India and Pakistan, there were many reports that war was about to break out, that each was about to attack the other. Yet these reports were baseless. The intelligence services in India and Pakistan are highly politicized. Information is not cross-checked by staffs the way it would be in the United States. The intelligence services often are in the dark about what their political leaders are doing. In India, Pakistan, North Korea, Iraq, and Iran, leaders are not likely to share information with their military forces because it might weaken their ability to control them.

Countries will spend money for bombs and missiles, but little for command control. Badly designed command and control systems will find "threats" where none exist. At the same time, they will not discern threats that do exist. As a result, attacks are actually invited because forces are sent to the wrong place or left unprepared. A poorly organized system makes countries feel insecure—and in fact they *are* insecure.

For example, when North Korea launched commando operations against the South, no one bothered to put the North Korean military on alert. When infiltration teams tried to assassinate South Korean leaders in the Seoul White House in 1968, and actually did so in Burma in 1983, the North Korean army gave no indication of realizing that its own side might be about to start a war. In a nuclear crisis, such ignorance could be highly dangerous if military units remained in vulnerable encampments where they could be struck with missiles. From the point of view of Seoul and Washington, that might seem to be a lucky break, but it would be more likely to drive the North toward a hair-trigger military response once its commanders understood their situation.

In one crisis caused by a large Indian army exercise in 1987, Pakistan intercepted radio reports suggesting that India was using the drill to conceal a real invasion. Whether such a radio transmission was authentic or garbled by Pakistani intelligence will never be known. What is known is that Pakistani leaders took the report quite seriously and thought war was imminent.

A lack of critical information is also a pervasive problem. Remarkably, in the same 1987 crisis, the Pakistani general in charge of defense moved his army out of its peacetime barracks to cover the border with India. He executed this maneuver in a fashion intended to maximize the chances that it would be picked up by Indian reconnaissance aircraft. The general's idea was to let the Indians know that he was prepared to block their invasion, so that they would call it off. Although the invasion did not occur, the Indians didn't notice this defensive repositioning for over two weeks! It isn't known whether their reconnaissance aircraft failed to spot it or they saw it but the information was never reported through the right bureaucratic channels. But it is a glaring example of what can happen when two half-modernized armies move toward each other. Had nuclear weapons been available to Pakistani forces, the Indian troops would have made a highly concentrated target at the border.

These situations are often worsened by cheap talk. With atom bombs to back them up, governments have a tendency to say things that aggravate a crisis. The same thing happened early on in the cold war, for example, when Khrushchev threatened "to bury the United States." But the dangers of such explosive rhetoric soon became apparent. By the end of the cold war the United States and the Soviet Union were using only the most restrained diplomatic language, even when their military moves were much more aggressive.

In Asian crises this lesson hasn't yet been learned. In the 1987 crisis between India and Pakistan, an Indian journalist was told by a senior Pakistani official that Pakistan would use an atomic bomb against India if its survival was endangered. The report appeared a month later in an Indian newspaper and, predictably, greatly increased tensions.

The shaky control of Asian nuclear forces increases the danger of accidental or unintended war. Asian states share all the problems of physical security and reliable communications that plagued the first nuclear states, but unique conditions in Asia heighten the dangers. The United States and the Soviet Union had confidence in their armed forces. Their commanders would not have moved recklessly forward without orders or with cavalier indifference to the meaning of those orders. Trust and understanding between civilians and the military is generally much lower in Asia than in the West. The level of mistrust is extreme to the point of pathology in North Korea and Iraq, but it also characterizes civil-military relations in India, Pakistan, and Iran. The lack of trust is compounded by deep mutual suspicion, by a history of plotting and coups, by ethnic and religious schisms, and by a military rife with treachery and ambition. These vary from country to country, but it seems fair to say that the world has reason to worry about whose fingers are on the nuclear triggers in countries such as Iran, Iraq, Syria, Pakistan, and North Korea.

Transitions of power could be especially dangerous. At these times there are generally no clear lines of authority. Chains of command that may be ambiguous under the best of conditions become even more obscure during a transition. In India the assassinations of two prime ministers set a dangerous precedent, suggesting one possible way to paralyze the command system, possibly blunting any return blow. In nations where the military has demonstrated recklessness, such as Syria and Iran, only the civilian leadership holds them in check. An upheaval in the government could open the way to military adventures with catastrophic consequence.

Problems in Communication and Bargaining with the West

In a crisis, the ability of the United States to influence and restrain events will be limited by fundamental cultural differences. This was a hard problem even during the cold war, but it looks to be even more difficult in the future.

Often one or both of the parties to a crisis—or at least elements within their leaderships—will want to exercise restraint but needs outside encouragement to prevail. In such a case, the United States must know how to signal its own interest in such restraint. The historical record shows that this is not a simple matter of exchanging official communiqués, either public or private. The United States has not been good at communication and bargaining outside of its own culture in the past. In 1945 it demanded that the Japanese surrender include the removal of the emperor. This demand created a major obstacle to ending the war. The United States failed to comprehend just how important this point was to the Japanese. Yet all official communications were limited to the demand for unconditional surrender, which precluded other attempts at bargaining. When the United States finally changed its demand, allowing the emperor to remain in power, it became clear what a small concession it had been for Washington to make in the interests of ending the war more quickly.

Communication problems were dangerously mishandled in the

Gulf War. The United States was in the dark about a threatened Iraqi escalation if coalition forces attempted to unseat Saddam's regime. Though informed about this possibility by Israeli intelligence, Washington never absorbed what was going on. On February 25, 1991, Iraq fired a missile with a concrete warhead into Israel's Negev Desert. This "blank" had the weight and speed profile of a biological warhead. It was believed by some experts in Israeli intelligence to be a signal that Saddam Hussein would unleash his germ arsenal against Israel if the allies invaded Baghdad to replace his regime. Yet even the Israelis remained unsure of what it meant. The remarkable feature of this event is that a weapon of mass destruction was used to tacitly communicate an escalation threat, but Washington never comprehended that this was the case. This is understandable, however, given the rapid pace of the war and the fact that Iraq and the United States had not developed the elaborate understandings that emerged between the two superpowers in the cold war.

Still, that both World War II and the Gulf War saw the United States having trouble in establishing good communication and bargaining understandings with its enemy suggests that the United States experiences major barriers to good crisis diplomacy, even if it has an interest in good communication.

Trading Off the Army

The heady excitement of possessing weapons of mass destruction, missiles, and other disruptive technologies has another downside: it leads impoverished countries to neglect their conventional military forces. Governments may skimp on their armies even more than they already do, in the mistaken view that such armies represent the old way of war, or simply because the new weapons are cheaper. The United States did this at the start of the cold war, badly neglecting the army until the policy was reversed in the 1960s.

Such neglect is already taking hold in Asia. Pakistan is increasingly reliant on its nuclear deterrent because economic chaos has made it impossible to modernize its army. In India and China the infantry is treated indifferently by civilian leaders. In South Korea the navy has received greater defense budget increases than the army, reversing decades of army domination over the budget.

Israel is even showing signs of such neglect. In the past, the air force and army were the most important strategic arms. Now Israel's inner security zone—Egypt, the West Bank, and Syria—appears reasonably secure. But Israel faces new challenges from long-range attack by Iran and Iraq. By enlarging Israeli security space to encompass the entire Middle East, it has imposed new trade-offs in Israeli defense.

Deterrence of a ground attack by Egypt and Syria is different from deterring a missile strike by Iraq and Iran. The air force and army lack the range and weapons to attack these countries now that their programs are dispersed and hidden underground. Israeli defense planners are debating whether an attack should be met with an all-out annihilating strike in return or a calibrated response of gradual escalation. A strategy of restraint needs to be communicated in advance if it is to evoke limitation in attack for fear of the consequences. But sending signals might be interpreted by enemies as a sign of weakness, most especially in the Middle East. These are difficult questions, with psychological as well as strategic implications. Israel's shift of its limited defense resources away from the air force and army—bedrock institutions that have intimidated its enemies for fifty years—is a risky change. The Israeli air force and army are among the most admired military forces in the world. Tampering with them to create a new strategic arm jeopardizes the intangible respect they command—a deterrent to adversaries in and of itself. Syria and Iraq still field large armies and could threaten Israel if it diverted too much attention and resources to long-range deterrence.

The cost of a submarine armed with nuclear missiles could ultimately reduce Israel's security.

On the other hand, failing to come to grips with the new threats endangers the very existence of Israel. This debate became public when officers who advocated a new strategic approach were passed over for promotion by commanders who felt they were not sufficiently supportive of the ground force.

On the other side of Asia, South Korea has concentrated on its army and eschewed missiles and weapons of mass destruction as the price of its alliance with the United States. Washington has used its position in South Korea to stop proliferation there, a move Seoul once seriously considered. But as North Korea has developed missiles, germ warfare, and possibly nuclear weapons, South Korea sees little benefit from investing more money in its army. Beyond the Korean Peninsula, Seoul sees a geopolitical landscape populated by giants—China and Japan—and seeks to build a military force relevant to that world. This view not only necessitates an expanded navy but also puts pressure on South Korea to build weapons of mass destruction of its own.

Terror and the Bomb

Asia is rife with sectarian disputes, which are likely to take on a more ominous character in an environment of weapons of mass destruction. The Babri Mosque incident in 1992 was touched off by Hindu zealots who demolished a sixteenth-century Mughal temple in northern India—supposedly built on a sacred Hindu site. The Hindu-Muslim violence that followed killed more than one thousand people throughout India and bolstered a dangerous fundamentalist streak in Indian politics.

Such incidents invariably polarize political tendencies. What Westerners find difficult to understand is the intensity of the feelings that Asians bring to these religious and ethnic disputes. Internal disorders could quickly spill over into whole regions, inflamed by mass media that reach across borders and by the political logic that seeks a foreign scapegoat for domestic problems. National leaders could then be backed into a rhetorical corner—a dangerous place for people who have atom bombs at their disposal.

In the Mideast the long-standing grievances of many terrorist groups against Israel and the United States have led to spectacular terrorist acts. The potential for non-state terrorism could go up sharply as well. Biological weapons, in particular, are much easier to get than nuclear ones.

Internal disorders can bring out the worst kinds of opportunistic behavior, by not only a country's own politicians but also neighbors seeking to egg on the internal combatants. It is easy to imagine that riots in the sacred cities of Mecca and Medina in Saudi Arabia could bring in Iraq or Iran with weapons of mass destruction. Either could intervene with a new kind of terror diplomacy—for example, issuing a nuclear ultimatum to Saudi Arabia, seeking to fracture its control over its Islamic population. Considering the potential for disorders to spill across borders, and the way nationalistic and religious battles have often moved beyond the point where a settlement can be negotiated, routine confrontations could escalate into cataclysm.

5   IS THERE AN EASTERN
WAY OF WAR?

The U.S. attack on Iraq in 1998 called attention to the vast difference in the weapons on which the two countries rely. Iraq's chemical and biological arms, the main targets of the raid, were indiscriminate weapons, meant to annihilate whole cities. The 400 cruise missiles fired by the United States were weapons of stealth and utmost precision. They could not be shot down because they could not even be seen by the defenders. They destroyed individual buildings, while barely breaking the windows of those across the street.

Strategically, moreover, the Iraqi regime has exhibited almost pathological opportunism, attacking Iran when it was weak, and Kuwait simply to steal its oil. There were no second thoughts about wiping out ethnic minorities in the most horrible way, or about planning mass civilian attacks against Israel. The United States, by contrast, avoided certain targets so as not to kill even a few innocent people. It ended the Gulf War as soon as it did, leaving Saddam in power, for fear of causing unnecessary casualties among the fleeing Iraqi troops.

An enormous change in attitudes toward war is under way, in both Asia and the West. As Asian countries are beginning to integrate nuclear weapons into their arsenals and strategic doctrines, Western countries increasingly are repulsed by them, and by war itself. Western governments display a growing distaste for the use of force and a horror of casualties, especially from direct combat. They have redefined their missions as "peacekeeping," not waging war. In the East, meanwhile, the tradition of guerrilla warfare is being abandoned because new technologies make it possible to attack the centers of Western power. These differences matter in thinking about the global balance of power. Even as the West retains its immense lead in technology and wealth, its real power is being curtailed by cultural and political forces at work within local societies.

STRATEGIC CULTURES: EAST MEETS WEST

The persistent patterns in the way nations fight reflect their cultural and historical traditions and deeply rooted attitudes that collectively make up their *strategic culture*. These patterns provide insights that go beyond what can be learned just by comparing armaments and divisions. In the Vietnam War the strategic tradition of the United States called for forcing the enemy to fight a massed battle in an open area, where superior American weapons would prevail. The United States was trying to refight World War II in the jungles of Southeast Asia, against an enemy with no intention of doing so.

Some British military historians describe the Asian way of war as one of indirect attacks, avoiding frontal assaults meant to overpower an opponent. This traces back to Asian history and geography: the great distances and harsh terrain have often made it diffi-

cult to execute the sort of open-field clashes allowed by the flat terrain and relatively compact size of Europe. A very different strategic tradition arose in Asia.

The bow and arrow were metaphors for an Eastern way of war. By
its nature, the arrow is an indirect weapon. Fired from a distance of
hundreds of yards, it does not necessitate immediate physical contact
with the enemy. Thus, it can be fired from hidden positions. When
fired from behind a ridge, the barrage seems to come out of nowhere,
taking the enemy by surprise. The tradition of this kind of fighting is
captured in the classical strategic writings of the East. The 2,000
years' worth of Chinese writings on war constitutes the most subtle
writings on the subject in any language. Not until Clausewitz, writing in the nineteenth century, did the West produce a strategic theorist to match the sophistication of Sun-tzu, whose *Art of War* was
written 2,300 years earlier.

In Sun-tzu and other Chinese writings, the highest achievement
of arms is to defeat an adversary without fighting. He wrote: "To win
one hundred victories in one hundred battles is not the acme of skill.
To subdue the enemy without fighting is the supreme excellence."
Actual combat is just one among many means toward the goal of
subduing an adversary. War contains too many surprises to be a first
resort. It can lead to ruinous losses, as has been seen time and again.
It can have the unwanted effect of inspiring heroic efforts in an
enemy, as the United States learned in Vietnam, and as the Japanese
found out after Pearl Harbor.

Aware of the uncertainties of a military campaign, Sun-tzu
advocated war only after the most thorough preparations. Even
then it should be quick and clean. Ideally, the army is just an
instrument to deal the final blow to an enemy already weakened
by isolation, poor morale, and disunity. Ever since Sun-tzu, the
Chinese have been seen as masters of subtlety who take measured
actions to manipulate an adversary without his knowledge. The

dividing line between war and peace can be obscure. Low-level violence often is the backdrop to a larger strategic campaign. The unwitting victim, focused on the day-to-day events, never realizes what's happening to him until it's too late. History holds many examples. The Viet Cong lured French and U.S. infantry deep into the jungle, weakening their morale over several years. The mobile army of the United States was designed to fight on the plains of Europe, where it could quickly move unhindered from one spot to the next. The jungle did more than make quick movement impossible; broken down into smaller units and scattered in isolated fire bases, U.S. forces were deprived of the feeling of support and protection that ordinarily comes from being part of a big army.

The isolation of U.S. troops in Vietnam was not just a logistical detail, something that could be overcome by, for instance, bringing in reinforcements by helicopter. In a big army reinforcements are readily available. It was Napoleon who realized the extraordinary effects on morale that come from being part of a larger formation. Just the knowledge of it lowers the soldier's fear and increases his aggressiveness. But in the jungle and on isolated bases, this feeling was removed. The thick vegetation slowed down reinforcements and made it difficult to find stranded units. Soldiers felt they were on their own.

More important, by altering the way the war was fought, the Viet Cong stripped the United States of its belief in the inevitability of victory, as it had done to the French before them. Morale was high when these armies first went to Vietnam. Only after many years of debilitating and demoralizing fighting did Hanoi launch its decisive attacks, at Dienbienphu in 1954 and against Saigon in 1975. It should be recalled that in the final push to victory the North Vietnamese abandoned their jungle guerrilla tactics completely, committing their entire army of twenty divisions to pushing the South Vietnamese into collapse. This final battle, with the enemy's army all in one place, was the one that the United States had desperately wanted to fight in 1965. If Washington could have arranged such a battle, it could have annihilated the North Vietnamese army. But the North was not about to let this happen in 1965. When it did come out into the open in 1975, Washington had already withdrawn its forces and there was no possibility of reintervention.

The Japanese early in World War II used a modern form of the indirect attack, one that relied on stealth and surprise for its effect. At Pearl Harbor, in the Philippines, and in Southeast Asia, stealth and surprise were attained by sailing under radio silence so that the navy's movements could not be tracked. Moving troops aboard ships into Southeast Asia made it appear that the Japanese army was also "invisible." Attacks against Hawaii and Singapore seemed, to the American and British defenders, to come from nowhere. In Indonesia and the Philippines the Japanese attack was even faster than the German blitz against France in the West.

Mao's strategy of a "people's war" also embodied the indirect approach. His writings seem almost lifted from *The Art of War.* Stealth was achieved by hiding in villages and the countryside, mingling indistinguishably with the people. Even in more conventional battles he relied on stealth, moving 400,000 troops into North Korea in 1950 by marching them at night. During the day the Chinese forces holed up in villages, where American reconnaissance aircraft couldn't spot them. The resulting attack against the American Eighth Army was a stunning surprise and led to the longest tactical withdrawal in the history of the U.S. army.

The greatest military surprises in American history have all been in Asia. Surely there is something going on here beyond the purely technical difficulties of detecting enemy movements. Pearl Harbor, the Chinese intervention in Korea, and the Tet Offensive in Vietnam all came out of a tradition of surprise and stealth. U.S. technical intelligence—the location of enemy units and their movements—was greatly improved after each surprise, but with no noticeable improvement in the American ability to foresee or prepare for what would happen next. There is a cultural divide here, not just a technical one. Even when it was possible to track an army with intelligence satellites, as when Iraq invaded Kuwait or when Syria and Egypt attacked Israel, surprise was achieved. The United States was stunned by Iraq's attack on Kuwait even though it had satellite pictures of Iraqi troops massing at the border.

The exception that proves the point that cultural differences obscure the West's understanding of Asian behavior was the Soviet Union's 1979 invasion of Afghanistan. This was fully anticipated and understood in advance. There was no surprise because the United States understood Moscow's worldview and thinking. It could anticipate Soviet actions almost as well as the Soviets themselves, because the Soviet Union was really a Western country.

The difference between the Eastern and the Western way of war is striking. The West's great strategic writer, Clausewitz, linked war to politics, as did Sun-tzu. Both were opponents of militarism, of turning war over to the generals. But there all similarity ends. Clausewitz wrote that the way to achieve a larger political purpose is through destruction of the enemy's army. After observing Napoleon conquer Europe by smashing enemy armies to bits, Clausewitz made his famous remark in *On War* (1832) that combat is the continuation of politics by violent means. Morale and unity are important, but they should be harnessed for the ultimate battle. If the Eastern way of war is embodied by the stealthy archer, the metaphorical Western counterpart is the swordsman charging forward, seeking a decisive showdown, eager to administer the blow that will obliterate the enemy once and for all. In this view, war proceeds along a fixed course and occupies a finite extent of time, like a play in three acts with a beginning, a middle, and an end. The end, the final scene, decides the issue for good.

When things don't work out quite this way, the Western military mind feels tremendous frustration. Sun-tzu's great disciples, Mao Zedong and Ho Chi Minh, are respected in Asia for their clever use of indirection and deception to achieve an advantage over stronger adversaries. But in the West their approach is seen as underhanded and devious. To the American strategic mind, the Viet Cong guerrilla did not fight fairly. He should have come out into the open and fought like a man, instead of hiding in the jungle and sneaking around like a cat in the night. The West is equally frustrated that Saddam Hussein survived the Gulf War, despite being pinned down, awaiting the final blow that could have ended his regime. U.S. hesitation at the war's end went against the traditional strategy of ending with a clean and decisive victory. Infuriated by this outcome, Washington has responded by demonizing Saddam Hussein, greatly exaggerating his importance. Retrospectively, regret over the lost opportunity and frustration over the subsequent events have begun to tinge the Gulf War, America's great redeeming victory in Asia, with the sense of failure and a job left unfinished.

The Western way of war leads to gigantic clashes as each side tries to bring the maximum force to bear on the battlefield. How Clausewitz would have assessed the wars fought by his students is an open question, but Sun-tzu surely would have been appalled at the colossal European battles of the twentieth century: the Somme, Stalingrad, Normandy. These battles demonstrate one enduring tendency in the Western way of war: excess. The West goes too far; this is part of its tradition. By 1916, two years into World War I, the consequences of achieving victory had become much worse for all of the participants than the peace they could have achieved by capitulating in 1914. Yet the dreadful slaughter continued for two more years. The Western penchant for excess in war is not to be forgotten, even now when it seems muted. In the cold war each side built more than 70,000 nuclear weapons, a number far surpassing any conceivable rational military use.

The strategic approaches of East and West have been described by Victor Hanson, a historian of antiquity, in the following way:

> The heavy infantry, the tactics of direct assault, and the very fire-power of American and European armies, which once captured the public imagination as somehow "heroic," have proven embarrassingly ineffective in the postcolonial conflicts and terrorist outbreaks of the years since the end of the Second World War. Instead, Western armies have become bogged down in the jungles and mountainous terrains of Africa, Latin America, and Southeast Asia, where they should never have been introduced, for both political and strategic reasons. The guerrillas and loosely organized irregular forces, the neoterrorists who for centuries have been despised by Western governments and identified with the ill-equipped, landless poor, now command attention, fear, even admiration. These forces have won respect not merely on political grounds, or even through any brilliance of combat, but rather because of their uncanny success at ambush and evasion of direct assault. They seek not to engage in but to avoid infantry battle.

CHANGES IN THE WAY OF WAR: EAST AND WEST

An extraordinary change in the way of war is under way. Asian countries have shed the traditions of guerrilla warfare that served them in their years of weakness. This gradual change, which began in the 1970s, is now abetted by weapons of mass destruction and missiles. A technocratic military that bets its future on weapons of mass destruction has replaced the valiant guerrillas and loosely organized irregular forces who stood up against the colonial powers.

As a result of this transformation, no longer are Asian forces restricted to stealthy attacks on isolated parts of Western forces that happen to be separated from their main units, hoping to catch them off guard to offset their technological inferiority. Direct attacks at the core of Western military power are now achievable. In the next Asian war the United States may find that its military bases and those of its allies—and soon even the U.S. homeland itself—are targets.

A significant part of Asia is moving into a modern era of war, leaving behind an agricultural past when the military was forged from masses of peasants. In embracing the weaponry of the West, the Eastern way of war is bound to change. The West will have to give up its romantic notion of stoic, courageous guerrillas and peasants as Asian countries develop modern bureaucratic militaries—inevitably characterized by the same kind of blunders, organizational obtuseness, and petty, opportunistic manipulations with which the West is all too familiar. A romantic strain in Western thought imagines Asia as technologically backward but endowed with honesty, simplicity, and purity of purpose—in contrast to the West's own sick fixation on power, domination, and technology. Asia is presumed to be in touch with different, older, and higher values.

Of course, it's easy to see that there is no Zen-like purity of purpose in North Korea's nuclear weapons or in Iraq's germ warfare regime. What is harder to recognize are the changes stemming from the introduction of missiles and mass destruction weapons into India and China. India is leaving behind its Gandhian tradition of nonviolence, offering up instead the tortured justification that its nuclear program will speed up the eventual global abolition of the bomb. But this argument only shows how self-generated military programs distort political language. China is leaving behind the egalitarian ideals once embodied by its People's Liberation Army. A Gandhian atom bomb or a People's ICBM are useful reminders that the West is not the only civilization capable of standing language and meaning on their heads in the pursuit of power.

A new way of war is also coming about in the West. A growing sensitivity to casualties makes it almost inconceivable that a Western country could fight a war like World War I again. Where is the officer brave enough to order a direct frontal assault on an enemy position, knowing that for every wounded soldier there will be a mother sobbing on CNN by nightfall? Instead of lives, the American military now expends money on stealth attacks by cruise missile and airplane, by precision strikes with laser-guided bombs, even by electronic warfare intended to paralyze the enemy's communications and control capability. All of these weapons kill at a distance and depend on pinpoint accuracy to minimize civilian casualties and damage.

A fantastic reversal is under way. Once the East embraced the stealthy techniques of the ambush, the archer, and the guerrilla as the only way to avoid direct, and therefore suicidal, clashes with superior Western firepower. Remarkably, the West now embraces stealth to lower its exposure to casualties. But it is stealth in a modern technological form. The U.S. cruise missile attacks on Afghanistan and Iraq in 1998 were like the ambushes and surprise attacks that once characterized Asian strategic culture. The adversary had no hint it was under attack until moments before the light flash from the missile's explosion. The missiles were launched from ships and submarines that crept to their firing positions without alerting the enemy.

Conceptually, this new Western way of war resembles the stealth tactics of the Asian archer, creeping up on a victim to loose a barrage. The difference, of course, is that modern Western stealth employs "arrows" that fire from 1,000 miles away and are far more deadly. They use rocket motors to propel their payloads of high-explosive fragmenting warheads over distances that Asia's bowman would have found unimaginable. The cruise missile has become the new weapon icon of the West, replacing the sword. The weapon of choice in the 1990s, it was first used in the Gulf War. Since then it has become the big stick of the United States in Asia, used three times since Iraq and Afghanistan through mid-1999.

Cruise missiles have no pilots. They are fired from distances of 1,000 miles from ships, submarines, and airplanes. Their slender profile makes them nearly invisible to radar, while the ships and planes that launch them can stand well off from their targets, undetected and safe.

Such long-distance strikes reflect a basic societal pressure to fight in a new way. No Americans were killed in any of these actions since the Gulf War because none was exposed to enemy fire. The emphasis on stealth armaments reflects a determination to go to almost any lengths to avoid casualties, using technology to destroy the enemy without getting close to him.

The experience of the U.S. army in Somalia offers a clear counterexample of the military and political dangers the new tactics hope to avoid. In 1993 lightly armed U.S. Rangers were isolated from their main units in downtown Mogadishu, Somalia's capital. The separated units called in reinforcements, but none could reach them in time to prevent their slaughter in a firefight against local militias. The U.S. soldiers' desecrated bodies were dragged through the streets amid the jubilant mobs. Gleeful, dancing soldiers and women and children brought to American living rooms a visceral taste of warfare such as they had never experienced. Popular opinion immediately coalesced around two options: either obliterate the guilty parties with massive force, which would have involved huge civilian casualties if it could have been accomplished at all, or quit the scene and return home. The collapse of the entire mission and the prompt withdrawal of all U.S. forces immediately followed.

U.S. stealth technology—the cruise missiles and stealthy aircraft—is driven by the need to avoid this kind of fighting. Stealth is thus more than the sum of its parts, more than a collection of technical innovations offering tactical advantages in combat. Radar-absorbing paints and oblique designs for missiles and airplanes provide not just invisibility but psychological distance. The force that destroys without being seen is the war machine of the society that cannot tolerate the visible killing of its own troops, or even those of the enemy. Stealth gives protection from enemy fire and from the disagreeable images of war. In historical terms, stealth technology appears to offer a way to remain the dominant military power in Asia without having to fight another Korea or Vietnam.

DR. STRANGELOVE, CALL YOUR OFFICE

It wasn't all that long ago that Americans loved their bombs, which guaranteed military advantage over a dangerous enemy, demonstrated technical prowess, and were a source of national pride. Film clips of American hydrogen bomb tests in the Pacific in the 1950s are today viewed with embarrassment. Forty years later we regret the vast environmental contamination that resulted from the production, storage, and testing of nuclear arms. But in the 1950s no one thought about the environment. The genuine menace posed by the Soviet Union and the need to deter an invasion of Western Europe are now easily forgotten. Stripped of the geopolitical context that made it seem necessary, the atomic bomb can seem like a huge mistake. But back then it was an important source of American pride. No one should be surprised to discover that Asian countries today feel the same way.

Western elites suffer a failure of imagination and memory. They cannot imagine how another nation could take pride in setting off a nuclear bomb, the weapon that just a decade ago seemed like a good bet to destroy civilization, and may yet do so. Forgetting their own history, they assume that any democratic government must share their anti-nuclear ethic. This is in the long tradition of Western colonial powers projecting their values onto Asia.

During the three weeks between the Indian and Pakistani atomic bomb tests, many Washington officials were certain that Pakistan would see the folly of following India down the nuclear path. They genuinely expected Islamabad to see the situation through the eyes of the West, which universally condemned the spread of nuclear weapons. Unfortunately, Pakistan did not see it this way and by all accounts never seriously considered anything other than responding in kind.

Just as the West sees nuclear weapons as objects to be removed from global politics, locked in a box and hidden in a closet, the nuclear countries of Asia pursue them without guilt or embarrassment; if anything, they flaunt their bombs. North Korea fired a rocket that actually flew directly over Japan's biggest cities before it splashed down in the western Pacific. In an accident even this unarmed test missile could have caused significant damage if it had fallen into a densely populated area. Yet North Korea seemed indifferent to the possibility. India and Pakistan detonated not single bombs but a series of five and six each, respectively.

In the cold war neither the United States nor even the Soviet Union ever flaunted their weapons in this manner. Doing so would have caused a storm of controversy. Many of the physicists who worked on the American nuclear program later felt extraordinary guilt over their involvement, and many of the best ones joined disarmament campaigns. One finds none of this attitude in China, India, Pakistan, or Iran. Asian scientists take pride in their accomplishments and are ambitious for the status that comes from participating in important government work. Of course, American nuclear scientists were also the only ones to build a bomb that was actually used against people. So far.

Democratic Israel has even been able to build hundreds of

bombs despite significant opposition from many civilian scientists and public figures. While opposition exists, successive governments ever since the 1960s have concluded that the bomb is necessary for security, and there appears to be little prospect of Israel's giving it up.

When Western arms control experts discuss nuclear weapons with military and government officials in Asia, they often come away quite puzzled. The experts are searching for carefully thought out rationales for how atomic bombs bolster national security, or any other evidence that the nuclear deterrence doctrines of the West are being adopted in some form in Asia. They are always disappointed. Broadly speaking, nuclear arms in Asia are not developed according to carefully conceived rationales, nor are they (so far) worked into detailed military plans. Rather, they are acquired without much thought for how they would actually be used. Asian countries want bombs, in part, just to have them, in the way that developing countries once wanted airlines as symbols of modernity.

Asia's long-standing technological inferiority is behind this impulse. The desire to catch up is rooted in nationalism, which in turn has a core of anti-Western sentiment dating back to the days when nationalism provided the unifying force to defeat Western colonialism.

In focusing on whether the West can keep its lead in technology, the United States is asking the wrong question. It overlooks the military advantages that accrue to societies with a less fastidious approach to violence. Weapons of mass destruction are still relatively new to Asia and have never been used against Western forces. Against an opponent like Iraq in 1991, one possessing a small germ warfare arsenal, U.S. conventional superiority was decisive, as the Gulf War clearly demonstrated. The willingness to intervene by massing forces on bases close to the enemy might have been different against an enemy that could reliably target these bases, or even the U.S. homeland.

The problem is not one of blowing things up on the battlefield. Rather, it is one of psychology and values. The new weapons can light up TV screens with images of war so frightening and awful that just their threat could keep the United States at bay. This would lead to gains without battle.

There are precedents for the strategic use of the psychology of fear, especially to shatter the morale of enemies that consider themselves more advanced. On the eastern front in World War II, combat degenerated into a barbaric struggle waged at the small-unit infantry level. In the battle of Stalingrad, after the Germans were lured into urban warfare, the two armies became so entwined that German airpower and artillery could not be used without firing on German troops. The German edge in weaponry—they, too, used technology to enable them to kill at a distance—was lost in the twisted canyons of the wrecked city. The great German strengths in planning, organizing, and supplying a modern war machine were useless in this primitive arena. The Viet Cong knew exactly what they were doing when they drove up U.S. casualties during the Vietnam War. They were trying to strip the United States of its technology advantage, which then lay in firepower and fast helicopter gunships. When the U.S. doctrine of counterinsurgency combined with airpower failed to work, U.S. morale sank. It was not just that the war was going badly for Washington, but that the enemy had transformed it into something that was, by American standards, primitive, barbaric, and frightening. Hanoi effectively "demodernized" the war.

Lyndon Johnson confronted an impossible situation. He could escalate. But this would drive up U.S. casualties. The United States desperately sought to "remodernize" the war, to put it on a basis that would substitute technology and money for American lives. Fantastic efforts were undertaken to strip away jungle canopy that hid the Viet Cong so that airpower could be used. Firestorms were considered to burn down the jungle. Huge tractors called Rome plows cut it back. The most stunning scheme was to place giant mirrors in outer space to reflect sunlight onto the jungle twenty-four hours a day, killing off vegetation from excessive growth. None of these schemes worked, and the effort to bring the war back to a high level of technology failed.

In the Gulf War the Iraqis were unable to do what North Vietnam did. The war was fought on U.S. terms. But weapons of mass destruction could become a major way for a country to threaten to "demodernize" a war against the West. As Americans find such weapons increasingly unthinkable, many countries can be expected to find them attractive for precisely this reason. They need not necessarily be fired, although this is always a possibility. Merely brandishing weapons of mass destruction would convey the implicit threat of a nasty, ugly war. Countries might, for example, make it clear that they fully intended to fire against cities rather than military targets. This danger is likely to be reinforced in the Western mind by the primitive character of these arsenals. With inadequate command and control and poor missile accuracy, a logical question has to be asked. How else could these weapons be employed other than in attacks on cities? Perversely, countries could score psychological points and add to deterrence by the crude design of their forces.

A CLASH OF CIVILIZATIONS?

Value differences between East and West point to the prospect of an impending clash between the two. One influential global strategist, Samuel Huntington, foresees just this possibility, a war between "civilizations." The idea of such a clash has been highly influential as a way to conceptualize the post–cold war world. He argues in his book *The Clash of Civilizations* that following the end of the cold war civilizations are becoming newly important organizing forces in world politics, transcending the nation-state as the basic unit of the international system. The United States and Europe together make up a civilization; Asia comprises several, including the Islamic, Chinese, Hindu, and Japanese. Huntington argues that the big wars of the twenty-first century are likely to pit one civilization against another. Muslims, for example, might attack the West with nuclear arms, in reaction to the modernizing but upsetting forces introduced by Western civilization. Or China, armed with missiles, might demand special concessions from the West to right past wrongs.

But the spread of weapons of mass destruction actually reinforces the importance of the nation-state, at the expense of civilizations. The technologies that underlie both war and wealth are organized at the level not of civilizations but of individual countries. More basically, civilizations don't go to war. Countries do. Civilizations don't control armies. Countries do. That weapons of mass destruction in Asia are organized by nation-states makes civilizations of declining importance in the political organization of Asia.

There are many other problems with looking at international relations in the framework of "civilizations," especially in the military realm. Civilizations are geographic regions that share a culture, institutions, and beliefs about the world. The world today can be broken down into eight major civilizations: Western (Europe and the United States), Latin American, the southern half of Africa, Orthodox (Russia, the Ukraine, and the Balkans), Islamic, Chinese, Hindu, and Japanese. Each shares a culture and beliefs about the world. Sharing common institutions is another matter. If a civilization has a shared culture and beliefs, but not institutions, then we must look at where key institutions are developing. In Asia national institutions of economic growth and defense are rapidly changing and becoming more important.

There are no transnational military institutions. Not even starry-eyed proponents of the United Nations would see its military peacekeeping arm as a central actor in Asian security. It helps to establish order in Cambodia, but on larger matters, such as nuclear testing on the subcontinent, China's claim on Taiwan, or the dispute in Korea, it has virtually no role whatever.

In the past, certain institutions of the West were paramount, and they did lead to usually one-sided clashes between civilizations. It is not too much of stretch to view the British navy in the nineteenth century as the military arm of Western civilization. It was at the very least *one* of Western civilization's military arms. It was a key instrument in spreading a common culture around the globe, namely, British colonial culture. Where this took hold—India, most of Africa, and Southeast Asia—it produced common ways of looking at the world. It backstopped political and legal structures that today are taken for granted.

The U.S. military could also be viewed as an institution of Western civilization, the heir to the British navy. The greatest democratizing institution in world history was the U.S. army. Through its occupation policies it ushered in democratic rule in the Philippines, Imperial Japan, and Nazi Germany. It played an important role in turning South Korea democratic as well.

In the twentieth century Western civilization was challenged by Russia. It, too, had institutions that it imposed on part of the world, chiefly Eastern Europe, in the form of the Warsaw Pact and its economic counterparts. The Red Army was primarily the representative of the ideology of communism, not of the civilization of the Slavic peoples. In any case, the collapse of Russian power has removed it from Europe and from much of what formerly constituted the Soviet Union as well, such as Ukraine and Kazakhstan. Even where Russian power remains sovereign, as in the Far East, its institutional capacity to rule is in steep decline.

It is true that the Islamic world tends to respond to the West in concert—for example, in defending Islamic culture against the intrusions of modern society. For its part, U.S. foreign policy tends to lump many of the countries of the Middle East together. The Islamic world is often seen in the United States as a source of terrorism and dangerous fundamentalism that could spread to Turkey and the countries of Central Asia. This is a civilizational perspective.

If Huntington is right, the coming clashes between civilizations could be especially bloody. Conflicts about basic values are worse than those over territory or for economic gain. Interests can be negotiated, but values are another matter. Value differences between parties who believe that to compromise means giving up on fundamental principles tend to produce not only larger misunderstandings but more persistent ones.

Civilizations can share important cultural traditions and beliefs but still not constitute a new framework for international security. The focus of the military modernization of Asia is the state, not the civilization. Commentators speak freely of an Islamic bomb, which is a highly misleading characterization. It conflates the military power of countries that have gone to war against each other, and that are themselves the principal objects of the development of weapons of mass destruction in the first place. The state, not the civilization, is behind modern armaments. Military forces are scaled in size to the nation-state. State leaders are the ones with their fingers on the nuclear triggers. No one recognizes the leaders of a civilization because it is not clear who these people are. Over the past decade there have probably been at least 500 declarations of holy war, or jihad, against the West. Most have no effect because there are no underlying institutions apart from the nation-states that could put them into effect.

Iran hardly feels secure from Iraqi SCUDs armed with biological weapons. Quite the opposite. Yet both are part of Islamic civilization.

Iraq's weapons push Tehran to arm itself as a counter. The brass rule—do unto others as they do unto you—is stronger than the ties that unite a civilization. Iran has been a target of Iraqi weapons of mass destruction, suffering debilitating chemical strikes in its war with Baghdad. The Islamic world did not come to Iran's assistance, nor was there even much of an outpouring of sympathy when Iranian soldiers were slaughtered by Iraqi chemical weapons. The lesson for Iran was to depend on itself for security, and on no one else.

Likewise, South Korea hardly feels secure from the buildup of Chinese military power, although both belong to the same civilization. South Korea's naval expansion of the 1990s is driven by fear of China, as well as Japan. Both make Seoul reach out to an outside power, the United States, for security. The South Korean naval program was intended to link that country more closely to the U.S. Navy, in anticipation of the day when American ground forces leave the peninsula and South Korea is left alone to deal with Japan and China in what could be a very different international environment than the one of the 1990s.

The spread of the new weapons increases the importance of the nation-state. Pierre Gallois, the French strategist of the 1960s, often remarked in lectures that there are no true alliances in the nuclear age. What he meant was that alliances in the past had a certain breathing space: a country could pull out at the last minute if it needed to do so. Thus, joining a coalition was a conditional act that did not foreclose independent behavior, because the agreement could always be broken. In the nuclear age, however, a country that does not provide its own security faces the prospect that it will not have time to assemble a deterrent when it is needed. Given the extraordinary destructive power of these weapons, it was hard for Gallois to imagine that a country would risk its own people to protect another's. In the jargon of cold war nuclear logic, the question always asked was: Would the

United States risk New York to defend Paris against the Soviet Union?

Although for now no one is asking those kinds of questions about, say, Seoul, in a very few years several Asian powers are likely to have missiles that can reach the United States. Then the question will become salient once again. One effect of the proliferation of weapons of mass destruction may turn out to be that nations become less dependent on alliances, more distant and self-reliant, and less inclined to trust in the bonds of culture, religion, and ethnicity that bind the members of a civilization. Countries may feel the need to build their own arsenals because their allies' pledges to protect them ring hollow. The horrible consequences of war only make national survival more a matter for the nation-state. Absent some new international organization that takes on the responsibility of preventing nuclear war, the future seems likely to be characterized by stronger, not weaker, nation-states.

Progressive thinking in the West views the nation-state as an outdated institution. The symbolism of the state, with its flag and national day, its dangerous tendency to fall into nationalism and chauvinism, and its absurd habit of comparing itself to other countries in everything from sports to school test results, is seen as a throwback to the age that gave the Western world two great wars within twenty-five years.

But even if the state is an outdated institution—and in many respects it is—envisioning a new form of political institution to replace it is difficult. Because neither civilizations nor international organizations like the United Nations and the International Monetary Fund deploy military forces, they are at an insurmountable disadvantage compared with states, which do. The cost of this disadvantage is growing as more states arm themselves with the new weapons.

And it was, after all, states that created modern Asia, not civilizations. Asian civilization could not compete against the West in the military, technological, or (until recently) economic spheres. Traditional forms of technology and economics had a retarding effect on development, and still do in the Islamic world. The Asian response to colonial rule was to appeal to the core values associated with their civilizations. But time after time these values proved inadequate against Western colonialism. What followed, what liberated China, Indonesia, Vietnam, India, and even fabricated states such as Iraq, was the nationalism fostered by the nation-state. It was the nation-state, and nationalism, that broke the West's colonial hold. The broader culture and beliefs of Asian civilization, whether Chinese, Japanese, or Islamic, could not do the job alone.

Iraq, Iran, North Korea, China, India, and Pakistan could not have mounted their present-day military programs without the energies of the state to raise money and mobilize talent. The civilizations that produced these countries were incapable of undertaking large technical projects before they became nation-states. Saddam Hussein's great achievement in Iraq was to create an effective state where none had existed before, out of a multiethnic hodgepodge of tribes and religions. That this program was perverted to evil purposes should not detract from the accomplishment. India's economic and military development, and even the experiences of countries as disparate as North Korea and Iran, all testify to the growing power of the nation-state in Asia.

The typical Western reaction is to point out that many states in Asia are largely artificial creations, with arbitrary borders and made up of diverse populations with little in common. The artificial character of many states does create many problems, beyond doubt. Many dividing lines arose from ad hoc decisions by colonial administrators rather than from historical or natural groupings of people. Iraq's boundaries arose much more from Britain's oil interests in the 1930s than from any historical roots going back to ancient Persia.

Lumping together the different branches of Islam, together with the Kurds, in a single country makes no objective sense.

Yet remarkably, it has made no difference that many Asian states are artificial constructs and that the very concept of the nation-state is a Western one imposed on Asia by colonizers. Asian leaders have fully bought into the idea, and they defend their sovereignty and borders as if they had existed for thousands of years.

The lesson of Asia is that a glorious invented history is almost as good as a real one. The national energies that can be unleashed on behalf of an imaginary tradition are extraordinary. Iraq is divided by religions and ethnicity, but compared to the Iraq of the 1930s it is a highly efficient and cohesive entity. Civilization offers a convenient resource for contriving a national past. Sometimes this need to manufacture unity goes to absurd lengths. The shah of Iran and his peacock throne built enormous shrines to the magnificent Persian past, with which he had not even the most tenuous relationship. His "ancient" kingdom actually began in the 1920s when his father was installed by the British as a figurehead. North Korea's efforts to build a national culture are even less plausible, based on a collection of ideologies that are eccentric to the point of psychopathology. National self-reliance has been transformed into a bizarre cult of worship for the ruling Kim family. Its other hallmark is national starvation. People keep expecting the regime to collapse of its own colossal inefficiency and incompetence at the most fundamental economic level, but it hasn't happened yet, and there can be no doubt that North Koreans would defend their territory with the utmost zeal.

A clash of civilizations could become a self-fulfilling prophecy if it were widely accepted in the West. The tendency to see the Islamic world as a social and military unity is especially dangerous. But the dangers of the twenty-first century are more likely to arise from a world of sovereign states.

Rather than a war of the West against other civilizations, a much more likely scenario is of the West's being drawn into regional conflicts because of weaker states' desperate need for strong outside allies. The fragility of many Asian regimes has led to neither their reform nor the appearance of alternatives to them. Rather, it has led to the acquisition of more armaments to protect them.

6   ACROSS THE STRATEGIC DIVIDE
New Powers and the Old Order

As one century turns to the next, the world is passing through a major strategic divide. The West's military dominance is declining as its monopoly on missiles and weapons of mass destruction ends. It retains enormous technical superiority in armies, tanks, airplanes, and even in software and computers. But this superiority counts for less because Asian missiles and weapons of mass destruction can attack critical Western weaknesses that arise from geography and politics, not technology.

A sweeping change is occurring in the structure of international security, distinct from the particular ambitions of individual countries. The structural features are the capacities of countries in Asia to strike at a distance beyond their borders; to quickly escalate the potential for violence in a crisis; to manipulate the threat of nuclear attack for political benefit; and to undermine or actually destroy the key foundations of Western military power in Asia. These are ineluctable, long-term trends. The particular ambitions of any one country, by contrast, are a function of shifting domestic politics and international alliances and tensions. China may cooperate with the

United States to keep regional order in Asia, or it may attempt to eject the United States from its bases in East Asia. India may accept the stalemate in Kashmir, or it may press Pakistan into an expensive arms race that the latter can ill afford. Iran may open up to Western overtures, or it may retreat back into Islamic fundamentalism. These countries may choose some middle course, or they may do something entirely different and unexpected. No one knows. But all of the possibilities, positive and negative, will be played out on a new chessboard, one on which Asian countries can strike back at one another, as well as at any outside power that becomes involved.

It is difficult to recognize this change because the usual tendency is to focus on individual national policies rather than on structural changes in the international security environment. Despite the potential for war and crisis on the Korean Peninsula, in South Asia, and in the Middle East, Asia today is largely peaceful. This relative peace obscures the fact that these countries retain the option of belligerence, even war, and the means to carry it out at a high level of violence. Foreign policy in Asia must be carried out with this possibility in mind. The problem is far more subtle than one of war versus peace. On the new Asian chessboard, countries can do all sorts of things, short of war, to try to get their way. India may choose to send another wake-up call about its feeling of being treated like a second-class nuclear power. It can do this by ostentatiously breaking any of several arms control agreements it has tacitly accepted following its atom bomb tests. Others can do the same. Countries that are now not even thought of as problems when it comes to weapons of mass destruction—Indonesia, Thailand, Australia—could be drawn into experimentation in this area, or even full-scale membership in the nuclear club.

Western leaders who fail to appreciate these structural changes—who still live in a world of slow-moving Asian armies incapable of reaching beyond their borders—will be surprised over and over. In the Gulf War the West was shocked by the missile attacks against Israel. Even though Washington knew that Iraq had such weapons, it didn't expect that nation to be reckless enough to fire them at Tel Aviv. After the war revelations about the scale of the Iraqi nuclear program and the extent of its preparations to fire germ weapons at Israel were even more surprising. In retrospect, it now seems clear that the United States seriously mistook the game Iraq was playing with its weapons of mass destruction, what it was willing to risk, and how it was willing to use force to pursue its interests.

The rise of Asian military power makes for a new relationship between the West and Asia. In the past Western powers could have prevailed in any showdown with Asian countries. In objective terms, the United States could have defeated China in the Korean War, or Hanoi in the Vietnam War, if it had been willing to pay the price in casualties and in world opinion, without endangering the U.S. homeland or those of its key allies. But this is changing as certain Asian states develop the capacity to strike back at American allies, at American bases in Asia, and soon, at the United States itself. The United States still retains overwhelming military power. North Korea could still be destroyed. But the United States can no longer do so without running the risk that it will take Seoul down with it, or Tokyo, or the thousands of American troops and civilians at Okinawa. Or soon, even Los Angeles.

This is what has changed in the world. The Gulf War marked a dividing line, even though Washington failed to grasp it at the time. Iraq's relatively primitive missiles could reach U.S. allies and bases, and it was prepared to arm them with biological warheads. This capacity and the will to use it change the American calculus enormously. The difference between the Viet Cong shelling the U.S. base at Pleiku with mortars in the Vietnam War and North Korea hitting Tokyo with anthrax bombs is immense. It would be suicidal for North Korea to launch such an attack, but the possibility that it might do so must make Washington tread more cautiously. This is easy to see by comparing how Iraq and North Korea are treated. Iraq, disarmed, is fair game for American cruise missiles and threats against its leadership, for failing to abide by arms control agreements. North Korea also fails to live up to its agreements but is treated far more gingerly.

The rise of Asian military power drives up the cost to the United States of remaining the world's only superpower, in several ways that are not obvious at first glance. One of the benefits of the end of the cold war was that neutral countries could no longer play off the United States against the Soviet Union to extract payments from each. India and Pakistan made a fine art of leaning just far enough toward Moscow to extract concessions from Washington, or vice versa, or sometimes both at the same time. In the future Asian countries will again be able to use this tactic by flirting with the acquisition of weapons of mass destruction rather than by courting Moscow. The great importance the West attaches to arms control—to the effort to make weapons of mass destruction "unthinkable"—creates a new form of leverage for countries that appear to be "thinking" about acquiring them. Leverage can be obtained in the form of direct support, as North Korea now receives from the United States, or it can be had indirectly in the form of increasing demand for sale of conventional weapons to substitute for nuclear weapons. It can be a demand for a security guarantee in which the United States is made responsible for the defense of countries as a bribe to prevent them from developing weapons of mass destruction for their defense. The costs to the United States may be limited only by the ingenuity of foreign leaders in formulating their demands.

There are also the direct costs. The United States can remain in Asia as long as it retains the ability to defend its bases and allies with antiballistic missiles. It can reduce its dependence on these bases in many ways and still be an important military power in

Asia. But it can't do any of these things cheaply. During the 1990s the United States was the world's dominant military power at the same time that it was cutting its defenses in response to the collapse of the Soviet Union. It was very easy to establish a domestic consensus around this policy because the benefits of being a superpower were considerable and its price was declining.

Raising the price of defense makes a national debate about American foreign policy much more likely, and also more complicated. An obvious example is the debate over national missile defense for the United States. This is under very serious consideration again, in response to the ballistic missiles of North Korea and other countries. In the early 1990s, however, national missile defense was presumed to be as dead as the cold war itself. Not only is it back as a major military and political issue, but it has become tied to the U.S. presence in Asia in ways that it never was in earlier debates over missile defense. The American public and Congress will never accept a theater missile defense that protects Japan and Saudi Arabia but not the United States. Otherwise, American citizens would be naked to attack from North Korea, Iraq, or Iran even as U.S. defense dollars were protecting the good citizens of Japan and Arabia.

The United States is encouraged by the rising costs of staying in Asia as a dominant force to search for "cheaper" strategies. One alternative is assembling groups of allies to share global responsibilities. Multinational cooperation is desirable, but it also imposes its own costs. In the showdowns with Iraq, the American-led coalition was perpetually "fraying" or even, when things got really tense, in danger of "unraveling." Many countries do have an interest in containing Iraq, but that is not their only interest. France and Russia look at U.S. actions in Iraq through the lens of their own agendas, which include preventing Iraq's nuclear armament but they also have other purposes. France seeks to limit the influence of the United States in order to increase its own status in the world. Russia tries to get away with exporting dangerous military technologies to Iran. It is difficult to manage this disparate coalition, but without it the costs of U.S. unilateral action to keep Iraq disarmed are prohibitive. For example, to enforce a trade embargo on Iraq against the wishes of Paris and Moscow would require that Washington challenge their flights into Baghdad—technically possible, but so costly in diplomatic terms as to be unfeasible.

The costs of staying a dominant military power are going up. After many years of a declining defense budget, calls in the United States for an increase in that budget have met with bipartisan support. Whether this support continues for any significant period of time remains to be seen, and whether any new expenditures are wisely spent is also a matter of uncertainty. In Europe the reaction to the rise of the Asian military power has been just the opposite: Europe has been priced out of the game. European governments have not been very interested in recent American proposals that NATO focus on dealing with weapons of mass destruction. Their thinking is that the possible solutions are so expensive—antiballistic missile defenses, space reconnaissance systems, and mobile armed forces— that there is little they can practically contribute to the cause. By default, most Europeans view the rise of Asian military power as an American problem, not a European one. Much more than in the cold war, the United States has very few allies it can turn to for anything more than moral and psychological support.

The United States does have alternatives as it confronts the increase in Asian military power, but its past strategy, maintaining overwhelming *conventional* military superiority, isn't likely to work as well in the future. One alternative is to seek variations on deterrence, otherwise known as brinkmanship. If China threatens to attack Taiwan, the United States can threaten to attack China. This surely would have important deterring effects. China or any other country in Asia does not lightly enter a military confrontation with the

United States. Two can play the game of manipulating the risks of war with weapons of mass destruction.

But having to resort to this strategy is itself very telling. It clearly shows the loss of Western dominance. There is an enormous difference for a nation between achieving its aims through unilateral military dominance in conventional military power and doing so through brinkmanship. Brinkmanship relies on the possibility of escalation to achieve those aims or to resist the other side's demands. Unpredictability is built into this strategy. When the military outcome is certain, deterrence is easy. But when it is not, doubt and hesitation creep into decisions. In addition, when a democratic society weighs the risks of a foreign attack, it can never forget that the potential casualties also vote. That is a problem that Iran, China, and North Korea don't have. The most enduring metaphor of the cold war was of each of the two superpowers holding a gun to the other's head. Now picture that same relationship with Beijing, Baghdad, Pyongyang, or Tehran, and you'll have an idea of one possible future.

Brinkmanship may be the best available strategy for dealing with some countries, but the rise in Asian military power is driving the United States to it. It is not a strategy consciously chosen by Washington because it is so risky. Sometimes there may be no alternative, but it is fraught with dangers of a kind that were thought to be safely buried with the cold war. NATO and Soviet forces faced each other across narrow borders for nearly forty years and managed to avoid catastrophe, so it can be done. But it required exceptional vigilance over the military, elaborate fail-safe systems, and a fantastically expensive worldwide surveillance and radar network. This is nobody's idea of what the benevolent post–cold war era was supposed to be like.

Exactly how quickly Asian military power will increase is a mystery, the sort of question that keeps think tanks churning. A wide range of estimates have been made, but in recent years the tendency has been to underestimate the speed with which Asian military power is rising. But whether the effects are felt in two years or ten, a corner has been turned. The countries of Asia are not going to reverse course and dismantle their new capability, not with any degree of Western pressure. Barring a general disarmament, the ones who gave up their weapons would be at a decisive disadvantage. And that's a problem not just for the countries of Asia but for the rest of the world as well. If one country were to decide to give up its weapons of mass destruction, regional order could be destabilized. For example, if Pakistan were to give up its missiles and A-bombs, India might well take advantage—pressing hard on the Kashmir front, for example. Facing Iraq and Iran, if Israel gave up its nuclear arsenal, the result would most likely be Middle East war, not peace. Except in the most abstract and idealistic terms, it's hard even to make a case that Israel should give up these weapons.

In some respects, the spread of missiles and weapons of mass destruction in Asia is like the spread of the six-shooter in the American Old West. The six-shooter was cheap and deadly. Known as the "equalizer," it leveled the fighting strength between men, making it a prime example of a disruptive technology, one that undermined the existing advantages of the powerful actors. The physically large and strong man now had to be cautious; his advantages had been rendered less important. With a six-shooter, a man's size and strength didn't matter. Because the weapon worked at a distance, the basis of dominance was changed. It is true that the six-shooter could be topped by better weapons. But few people wanted to go to the expense and trouble of obtaining a Gatling gun to gain an advantage over the six-shooter. Instead, the six-shooter changed the power balance in the West permanently.

WESTERN RESPONSES

It is useful to consider some other possible responses to the rise of Asian military power—even ones that no one would seriously undertake. Looking at them in pure form helps clarify the benefits and disadvantages of each. Some may seem extreme and outside the realm of political feasibility. They are nonetheless useful to mark the borders of what sort of actions are acceptable and desirable.

Preventive War

In the early 1950s Bertrand Russell called on the world to disarm the Soviet Union forcibly before it acquired a large stock of atomic bombs. One possibility in the current situation is to use military force to counter the spread of the bomb. This option has been contemplated on a number of occasions in history. In the summer of 1964, with China set to test its first atomic bomb, the Johnson administration drew up plans for U.S. air strikes against the Chinese production and testing facilities at Lop Nor, deep in western China. Both conventional and atomic bombing were contemplated. McGeorge Bundy, President Johnson's national security adviser, at first favored a preemptive attack, and he even approached the Soviets to get their reaction.

The Soviets would have none of it. The idea was shelved as both risky and counterproductive, as well as vaguely un-American in its sneakiness. Even without its nuclear weapons, China could still reignite war in Korea or intervene directly in Vietnam. Ultimately President Johnson and Secretary of Defense McNamara decided that strikes would only reinvigorate the Chinese bomb program and, by showing how much the United States feared them, accelerate the spread of atomic weapons elsewhere in Asia.

In 1969, when tensions mounted between Moscow and Beijing in the Russian Far East, the tables were turned. The Soviets now approached the United States about its willingness to countenance Soviet strikes to take out the Chinese nuclear capacity. This time Washington said no. The Chinese bomb, and the Western failure to stop it, reaped major benefits for Beijing. Washington and Moscow treated China with far more respect after it went nuclear.

In June 1981 the Israelis did what the United States and the Soviet Union would not. They launched air strikes against an Iraqi nuclear reactor located at Tuwaitah. The French-built Osirak research reactor could have produced enriched uranium and small quantities of plutonium for Saddam Hussein's first atomic bomb program. After a wrenching discussion in the cabinet about the risks, Prime Minister Menachem Begin ordered the strike that successfully destroyed the reactor.

However, the attack provoked a much greater effort by Iraq to conceal its program by dispersing it to underground sites beyond the range of Israeli aircraft. There is reason to believe that the Osirak raid also led Iran, North Korea, and India to conceal their bomb programs, making them much harder to attack.

The question arises: Will the West launch military strikes on countries to prevent them from getting weapons of mass destruction? This is an extreme measure, of course. But many sober experts have adopted in essence Bertrand Russell's argument. The danger of proliferation is clearly one of the highest priorities of the U.S. government, and military strikes have repeatedly been launched against Iraq to prevent it from recovering its earlier capabilities.

But overall, as a general Western policy with widespread support, preventive war's disadvantages outweigh its advantages. The only reason that a kind of slow-motion preventive war is waged against Iraq is because Baghdad was opened up for inspection following its loss in the Gulf War. There have been no military strikes against North Korea, Pakistan, India, or Iran. They all have the ability to respond with major military counteractions of their own, and faced with this possibility, the West finds that its enthusiasm for preventive war virtually disappears.

The character of proliferation also makes preventive war a very difficult technical undertaking. The problem is that facilities for missile construction and biological and chemical weapons are very hard to find. The United States cannot carpet-bomb large areas of suspect countries. As missiles and bombs spread through Asia, the technical problem becomes ever more daunting because the list of possible targets grows to levels that make attack impractical.

Arms Control

Arms control, by formal treaty, habit, and moral suasion, is an important feature of the West's response to the rise of Asian military power. It is needed to legitimize international efforts to check the spread of the bomb. Without arms control as an objective, the West simply looks too much as if it wants to keep its monopoly on advanced military technologies that could upset its position in the world. With arms control, this desire can be put in a broader context of purposes that are in the broader interest of mankind.

The Western world's emphasis on arms control also encourages violator countries to keep their weapons programs under wraps. If the moral sanctions against weapons of mass destruction were ever lifted, countries would be forced to speed up their programs, since they would fear that their neighbors and "everybody else" were doing so. International tensions would be sure to heat up, and nuclear weapons would return to the center of world attention.

It is in the interests of the West to keep this from happening. The return of atomic weapons to the center of international politics would increase their status; if countries were openly to brandish them by making threats based on their use, they might in fact become more usable. All of this is fairly well understood. The problem with arms control is the tendency of negotiators to pursue it for its own sake, losing sight of the larger objective: the national security of the United States, and international order in the world.

Arms control faces many daunting challenges, so much so that it seems clear that some new kind of arms control is needed. Existing approaches have served well for several decades but are now not doing the job they once did. As the nuclear club increases in size, the likelihood grows that the club will expand even more, in what an engineer would call an example of "positive feedback." India gets the bomb, so Pakistan must follow suit. Those who do not officially join the club by testing a weapon must become increasingly suspicious as they watch these new developments. There is likely to be much more "cheating" in the future as countries hedge against the possibility that their neighbors are working undercover to build these weapons. Such a widespread increase in suspicions about what is going on behind the scenes conveys the impression that the whole arms control framework is falling apart. We need imaginative new proposals for coming to grips with these problems in ways that restore confidence in arms control's contributions to U.S. national security and international order.

Balance of Power

One response to rising Asian military power is to exploit the natural differences in interests between countries, to balance one off against the other. This is, of course, a classic strategy, and one used by all major world powers throughout history. In present-day terms, there are many examples of what a balance-of-power approach to Asia would be like. The Indian nuclear effort could be seen as a way to counter China's buildup. Pakistan's bombs could offset India's. In Northeast Asia the expansion of the Japanese and South Korean navies would strengthen the power of states bordering China. At the other end of Asia, the United States has always been happy to exploit the divisions in the Middle East between secular, Sunni, and Shiite states, conservative and radical governments, oil-rich and oil-poor countries, to forestall the emergence of a united Arab challenge to the American position there.

Balance-of-power approaches are controversial. For one thing, they can backfire, as they did when the United States supported Iraq in its war with Iran, strengthening a future enemy. But they have the virtue of being natural features of a world that divides itself according to nation-states. Rivalries between nation-states create a balance of power, or more frequently a balance of interests, without the need for continual American intervention of a kind that can be very expensive in political and economic terms. The Indian atomic bomb tests automatically created a situation wherein China was sandwiched in between India and Japan. China now has to factor into its arms decisions, and its foreign policy, a raft of constraints that now loom larger. Beijing has a potential competitor in India, and it cannot want another in Japan. For this reason, Beijing has less of a free hand in Asia than would otherwise be the case. The Chinese must take care to avoid actions that could spark larger defense programs in India and Japan, either of which would hurt China's focus on its own economic development.

The United States could readjust the balance of power between states by sharing its intelligence and technology—as it has long done with NATO allies and friendly states such as Israel. Indeed, it would be surprising if future U.S. technology-sharing to alter military power balances did not come to include many more countries in Asia.

World Government

So as to cover all the possibilities, world government must be considered as a way to avoid the problems raised by the spread of missiles and weapons of mass destruction (along with, for that matter, weapons of piecemeal destruction). Under a world government, states would not be permitted to acquire these weapons. Moreover, the role of force would be globalized and removed from the control of any one country; presumably it would be given to some world legislative or executive body.

Proposals for world government were often put forth in the 1950s as a way to cut the Gordian knot of the cold war. Not much came of them, and the idea was pretty much abandoned for decades. It has resurfaced in recent years in the less threatening form of proposals to strengthen the United Nations and other existing international bodies. The idea is to change the norms of international behavior through sanctions and penalties so that states actually have an incentive to abide by the UN's charter and the decisions of the Security Council. By establishing a tradition of turning to these groups to settle disputes, it is hoped that the world might make a gradual and relatively painless transition to an informal system of world government.

Unfortunately, the evidence from most parts of the world is that this isn't happening. Major doubts remain about the feasibility of world government, and continued tests of missiles and bombs in defiance of treaties and world opinion only make it seem less likely.

No Exit

There is another possible strategy: the United States could withdraw its military forces from overseas and let Asians work out their own security system for the future. Such a withdrawal especially makes sense to anyone who believes in an impending clash of civilizations. In such a world, a Western presence in Asia would be catalyzing, one that drew fire from united opposition against the values of the West. Ideally, each civilization would retreat to its core area and steer clear of adventures elsewhere.

But the sources of conflict in Asia arise from nation-states, not civilizations. Far from benefiting from a withdrawal, a pullback could prove disastrous for the United States, and for Asia. One problem is that civilizations are not united. Another is that the nature of the new military technology is such that a country can build up a significant advantage over its neighbor. The two clearest examples of this are today the backbones of U.S. military planning, in the Middle East and on the Korean Peninsula. A U.S. withdrawal, to let Saudi Arabia and Kuwait defend themselves against Iraq and Iran, would run the grievous risk that those countries would be taken over by the stronger powers. In East Asia, likewise, the United States cannot safely withdraw from the region without increasing the chances of war in Korea.

The rise of Asian military power does not argue for a U.S. pullback from the world militarily. It argues instead for a restructured U.S. military, one that can operate at greater distances from home and is less reliant on vulnerable forward bases. Bases may be preserved for symbolic reasons, if the costs are not too high. They also have great political significance: they reassure allies that the United States shares an interest in their defense. But the United States cannot continue to base its fighting power in these installations because they are becoming too vulnerable to attack.

Undertaking to protect them with theater missile defenses is a very expensive proposition, one that could lead to a strategic dead end for the United States. Putting protective bubbles around U.S. bases in Asia has a great deal of appeal in the foreign policy community, because it extends the stay of a visible American military commitment on the ground in Asia. But alternative options, such as restructuring U.S. forces so that they do not need forward bases in the first place, are overlooked in the desire to hang on to a form of presence in Asia that cannot possibly endure with any level of expenditure.

Instead, more of the U.S. military has to be based in the United States, and concentrated in mobile forces that can use stealth, speed, and deception to enhance their survivability. The United States already has this capability to a great extent in its aircraft carriers, submarines, and long-range bombers. It is not a major step to allocate more resources to these areas, and to research and development along these lines.

The United States can retain bases in Japan, South Korea, Guam, and the Middle East. But the value and the meaning of those bases is declining. In the future, overseas bases should be storage areas for low-value logistics or training centers. They could be made up of Quonset huts with a U.S. flag in front, a visible statement of commitment but not an attractive target.

A restructured U.S. military should be complemented by more attention to the Asian balance of power. One clear consequence of the rise of Asian military power is that the United States must reach out beyond the Western club of nations. The days when the only important U.S. allies were Britain and Germany are past. To be sure, using the word *ally* to describe India or China does not do justice to the complexity of their relationship with the United States. But the United States is going to need the help of a much wider range of countries than has been true for most of its highly Eurocentric history.

One good outcome for the United States would be to have regional superpowers that can play the role of regional arbiters. A world in which a handful of big powers like China and India are well armed with missiles and weapons of mass destruction to stabilize their regions is not nearly as bad as some of the alternatives. It would be far better than a mad scramble by everyone to acquire atom bombs. Looked at this way, Indian possession of nuclear arms is beneficial because it establishes that country as a major power and blocks China from trying to increase its influence in the region. It is possible that the United States would encourage the proliferation of weapons of mass destruction in Asia as a way to establish a better balance of power there. A nuclear-armed Japan—however unlikely that seems now, given its domestic politics—might have many advantages for the United States, representing an offsetting power over China that in many ways would be much more natural than it is for the United States to assume this role. This is by no means the international order envisioned in the West in the early 1990s, when there was essentially only one major power, the United States. But in coming years an international order based on regional superpowers is likely to look much less shocking than it does today.

Many problems would remain in a world of regional powers in Asia. In South Asia such an arrangement would virtually concede Indian domination of Pakistan. In the Far East it would give China a free hand to take Taiwan. A country with weapons of mass destruction could cause a great deal of damage to regional order. But the United States cannot be pushed in the opposite direction. It cannot come to the assistance of every smaller state that might be threatened by a larger power. One policy advanced in the wake of the Indian nuclear tests called for providing a U.S. security guarantee to Pakistan in exchange for a pledge that it would not join the nuclear club. That agreement would obligate the United States to come to Pakistan's defense if it were menaced by India, or anyone else. As it turned out, Pakistan went ahead and tested its bombs anyway, but there are some who continue to call for a wider use of such security guarantees. In the early 1990s, when U.S. power seemed invincible, a policy that required the United States to take on the responsibility for defending Taiwan and Pakistan, and perhaps others, against stronger powers might have seemed reasonable. But the dangers of such a policy have greatly increased, and security guarantees need much more careful consideration before they become a central part of U.S. foreign policy.

THE SPECIAL CASE OF JAPAN

Japan is an anomaly in Asia, and for that matter, in the world: a rich, technologically advanced nation with a small military establishment and no visible ambition to make it larger. Tokyo conceives of security more in economic than military terms. At the same time, the regional environment in East Asia is becoming more heavily armed. Asia's burgeoning economies are now supporting a military transformation to longer-range, more deadly weapons, as this book has argued. North Korea is the most aggressive threat to Japan, but China and South Korea are developing military postures that Tokyo can no longer ignore, as it did during the cold war. Japan's dependence on the Persian Gulf for oil makes it vulnerable to the military situation there as well.

It is very easy to forget just how unique Japan is. Japan was the first Asian country to challenge the West with the West's own technology. As the only country where atom bombs were dropped in warfare, it is profoundly opposed to nuclear weapons. Its defense spending is less than 1 percent of GDP, and a near-consensus national policy is that rearmament to provide for its own security is "unthinkable." Virtually every opinion poll in Japan shows almost no support for an increased defense effort.

It is very difficult to have a serious discussion about Japanese defense because of prevailing attitudes. Debates have an air of unreality to them. Japan's views on military and security matters resemble those common in the United States before World War I. Then, the British navy protected the United States. The U.S. view at the time was that Europe was fixated on war, always thinking about war and always preparing for it. This compulsion itself was thought to produce the foreign wars that the United States up to that point had avoided. Militarism was attributed to the Europeans and declared to be "un-American," a feature of the national character that distinguished the United States from Europe. This view, of course, was fairly annoying to Europeans, who did not have the luxury of what the historian C. Vann Woodward has called "free security"—an accident of geography that allowed the United States more or less to ignore national defense until the twentieth century.

This is very much the attitude one still finds in Tokyo: there are no major military threats to Japan, and the other countries in Asia will inevitably follow Japan's economic model and in the process leave behind any thoughts of getting their way by force. Japan is content to choose between different kinds of economic security. One such choice that is often posed is whether Japan should become an Asian Sweden or Switzerland. In the Swedish model, Japan would finance international peacekeeping, environmental, and humanitarian efforts, while embracing a largely pacifist position on its own defense. In the Swiss model, Japan would withdraw from these activities and focus more narrowly on its own business interests.

Japan was instrumental in making the Pacific Basin a peaceful trading area, working in concert with the United States. Now competitors are appearing who have adapted the Japanese economic model but not its inhibitions about the military. As their growing wealth enables them to finance modern weapons, Japan is likely to find that its defense choices cannot be defined only in terms of economic approaches.

There is a certain irony, even a perversity to all of this. Japan's desire has been to prove by example the superiority of economic over military power. On its own terms, it has succeeded brilliantly. But the story is taking a different turn than expected. China and other Asian countries are not following the Japanese pacifist example, the way they followed its economic example. In the next century this dynamic may force Japan finally to escape the shadow of World War II. A more visible military program could well be required, to demonstrate to the world that the old approaches of exclusively economic security are no longer binding on its future behavior.

NEW POWERS AND THE OLD ORDER

The emergence of new technologically armed states, and their peaceful assimilation into the world community, is without doubt the central challenge of the twenty-first century. Until recently this challenge was defined in economic terms, and the question was whether China and others could be absorbed into the worldwide trading and monetary systems. This economic challenge remains, and it is daunting. But it now is complicated by a related military one.

It would be dangerous to give too much recognition to countries that have atomic bombs and missiles. A North Korea with atomic bombs is not a world power. But the greater danger comes from

The countries that must be brought into a global economy have substantially greater military capacities. It is therefore no longer possible to dictate high-handedly what the entry conditions are. The problem here is not one that is subject to technical fixes; rather, it entails a newly restructured international system in which the West's military superiority no longer goes unquestioned and its capacity to backstop its vision of world order is sharply limited compared to what it was even a few years ago.

ignoring the rise of Asian military power, of speaking in the esoteric jargon of nonproliferation strategies and not linking it to the broader economic transformation that has reshaped the Asian landscape in the last decades. It is one thing not to want to acknowledge the status of a nuclear crackpot state like North Korea. But it is another to deny the fundamental transformation at work: economic and military potentials have always been interlinked, and Asia's acknowledged economic rise is now leading to a military rise as well. That the most impoverished of all Asian states, North Korea, is part of a trend that includes the continent's most rapidly growing and dynamic country—China—and its democracies—India and Israel—reinforces the truly fundamental character of the forces at work.

At present those countries that industrialized first, Europe and the United States, retain their special position in the world. The world is still run by a Western club, and there is little use in denying it. International organizations like the United Nations and the World Bank have been instruments of Western influence. They were created by the victorious powers of World War II, and they have reflected the interests of the most developed countries. This system was backed in last resort by U.S. military power, which for a long time monopolized crucial military technologies. As Asian economies return to their upward climb, a new recognition of the changing power conditions in the world is inevitable. As their economic and military clout grows apace, the new powers are unlikely to tolerate the old order to anything like the extent to which they did in the postcolonial era, when they had no alternative but to do so. The rules of the old order will have to change to better reflect the interests of a large part of the world that has not received the recognition that its numbers, and power, demand. That France and Britain retain permanent seats on the UN Security Council while India and Indonesia, or for that matter Japan, do not, is, of course, ridiculous. This is but one example among many of an institution that has not changed to reflect the new conditions of Asia's significance.

The proper place in the world of Asia's ancient civilizations has not been recognized because Asian technology has been too backward. As a result, Asia has not yet had much experience in contributing to world order. Japan is the exception, but a unique one. The challenge to the West is not only one of managing a complicated integration of new countries into a world order that it has run for centuries. It is a challenge of self-conception. The challenge comes from the realization that the West's conception of itself is that of a leader that shapes international security and economic affairs. The long era in which Asia was penetrated by outside powers is coming to a close. An age of Western control is ending, and the challenge is not how to shape what is happening but how to adapt to it.

BIBLIOGRAPHY

This book is a synthesis of a wide range of scholarly and serious contemporary works. Principal sources for ideas are shown below. These books are an excellent reading list for those trying to develop new ways of understanding the extraordinary changes that are now taking place in the world.

Baudrillard, Jean. *The Gulf War Did Not Take Place.* Bloomington, Indiana: Indiana University Press, 1995.

Christensen, Clayton M. *The Innovator's Dilemma, When New Technologies Cause Great Firms to Fail.* Cambridge, Massachusetts: Harvard Business School Press, 1997.

Dunn, John, ed. *Contemporary Crisis of the Nation State?* Oxford: Blackwell, 1995.

Farmer, Edward L., et al. *Comparative History of Civilizations in Asia.* Volumes I and II. Boulder, Colorado: Westview Press, 1986.

Giddens, Anthony. *The Nation-State and Violence.* Berkeley, California: University of California Press, 1987.

———. *The Constitution of Society.* Berkeley, California: University of California Press, 1987.

Hanson, Victor Davis. *The Western Way of War*. New York: Oxford University Press, 1989.

Harries, Meirion and Susie. *Soldiers of the Sun, The Rise and Fall of the Imperial Japanese Army*. New York: Random House, 1991.

Haselkorn, Avigdor. *The Continuous Storm: Iraq, Poisonous Weapons, and Deterrence*. New Haven, Connecticut: Yale University Press, 1998.

Huntington, Samuel P. *The Clash of Civilizations and the Remaking of World Order*. New York: Touchstone, 1996.

Kahn, Herman. *Thinking About the Unthinkable*. New York: Horizon Press, 1962.

Kahn, Herman and Pepper, Thomas. *The Japanese Challenge: The Success and Failure of Economic Success*. New York: William Morrow, 1980.

Kedourie, Elie. *Nationalism*. Cambridge, Massachusetts: Blackwell, 1993.

Kundera, Milan. *Slowness*. New York: HarperCollins, 1995.

Lewis, Martin W. and Wigen, Karen B. *The Myth of Continents*. Berkeley, California: University of California Press, 1997.

McNeill, William. *The Pursuit of Power*. Chicago: University of Chicago Press, 1982.

Pannikar, K. M. *Asia and Western Dominance, A Survey of the Vasco da Gama Epoch of Asian History, 1498–1945*. London: G. Allen & Unwin, 1959.

Pfaff, William. *The Wrath of Nations: Civilization and the Furies of Nationalism*. New York: Simon & Schuster, 1993.

Ralston, David B. *Importing the European Army: The Introduction of European Military Techniques and Institutions into the Extra-European World, 1660–1914*. Chicago: University of Chicago Press, 1990.

Toffler, Alvin and Heidi. *War and Anti-War, Survival at the Dawn of the 21st Century*. Boston: Little, Brown and Company, 1993.

Woodward, C. Vann. *The Future of the Past*. New York: Oxford University Press, 1989.